Very Truly Yours, Nikola Tesla

Very Truly Yours, Nikola Tesla

by Nikola Tesla

Wilder Publications, LLC.
PO Box 3005
Radford VA 24143-3005

ISBN 10: 1-934451-91-6
ISBN 13: 978-1-934451-91-5

First Edition

10 9 8 7 6 5 4 3 2 1

Table of Contents

Mr. Nikola Tesla on Alternating Current Motors

Electrical World — N. Y. — May 25, 1889

To the Editor of The Electrical World:

Sir: About a year ago I had the pleasure of bringing before the American Institute of Electrical Engineers the results of some of my work on alternate current motors. They were received with the interest which novel ideas never fail to excite in scientific circles, and elicited considerable comment. With truly American generosity, for which, on my part, I am ever thankful, a great deal of praise through the columns of your esteemed paper and other journals has been bestowed upon the originator of the idea, in itself insignificant. At that time it was impossible for me to bring before the Institute other results in the same line of thought. Moreover, I did not think it probable — considering the novelty of the idea — that anybody else would be likely to pursue work in the same direction. By one of the most curious coincidences, however, Professor Ferraris not only came independently to the same theoretical results, but in a manner identical almost to the smallest detail. Far from being disappointed at being prevented from calling the discovery of the principle exclusively my own, I have been excessively pleased to see my views, which I had formed and carried out long before, confirmed by this eminent man, to whom I consider myself happy to be related in spirit, and toward whom, ever since the knowledge of the facts has reached me, I have entertained feelings of the most sincere sympathy and esteem. In his able essay Prof. Ferraris omitted to mention various other ways of accomplishing similar results, some of which have later been indicated by 0. B. Shallenberger, who some time before the publication of the results obtained by Prof. Ferraris and myself had utilized the principle in the construction of his now well known alternate current meter, and at a still later period by Prof. Elihu Thomson and Mr. M. J. Wightman.

Since the original publications, for obvious reasons, little has been made known in regard to the further progress of the invention; nevertheless the work of perfecting has been carried on indefatigably with all the intelligent help and means which a corporation almost unlimited in its resources could command, and marked progress has been made in every direction. It is therefore not surprising that many unacquainted with this fact, in expressing their views as to the results obtained, have grossly erred.

In your issue of May 4, I find a communication from the electricians of Ganz & Co., of Budapest, relating to certain results observed in recent experiments with a novel form of alternate current motor. I would have nothing to say in regard to this communication unless it were to sincerely congratulate these gentlemen on any good results which they may have obtained, but for the article, seemingly inspired by them, which appeared in the London Electrical Review of April 26, wherein certain erroneous views are indorsed and some radically false assertions made, which, though they may be quite unintentional, are such as to create prejudice and affect material interests.

As to the results presented, they not only do not show anything extraordinary, but are, in fact, considerably below some figures obtained with my motors a long time ago. The main stress being laid upon the proposition between the apparent and real energy supplied, or perhaps more directly, upon the ratio of the energy apparently supplied to, and the real energy developed by, the motor, I will here submit, with your permission, to your readers, the results respectively arrived at by these gentlemen and myself.

Energy apparently supplied in watts.		Work performed in watts.		Ratio of energy apparently supplied to the real energy developed.	
Ganz & Co.	Westinghouse Co.	Ganz & Co.	Westinghouse Co.	Ganz & Co.	Westinghouse Co.
18,000	21,840	11,000	17,595	0.611	0.805
24,200	30,295	14,600	25,365	0.603	0.836
29,800	43,624	22,700	36,915	0.761	0.816
......	56,800	......	48,675	.....	0.856
......	67,500	......	59,440	.....	0.88
......	79,100	......	67,365	.....	0.851

If we compare these figures we will find that the most favorable ratio in Ganz & Co's motor is 0.761, whereas in the Westinghouse, for about the same load, it is 0.836, while in other instances, as may be seen, it is still more favorable. Notwithstanding this, the conditions of the test were not such as to warrant the best possible results.

The factors upon which the apparent energy is mainly dependent could have been better determined by a proper construction of the motor and observance of certain conditions. In fact, with such a motor a current regulation may be obtained which, for all practical purposes, is as good as that of the direct current motors, and the only disadvantage, if it be one, is that when the motor is running without load the apparent energy cannot be reduced quite as low as might be desirable. For instance, in the case of this motor the smallest amount of apparent energy was about 3,000 watts, which is certainly not very much for a machine capable of developing 90 h. p. of work; besides, the amount could have been reduced very likely to 2,000 watts or less.

On the other hand, these motors possess the beautiful feature of maintaining an absolutely constant speed no matter how the load may vary. This feature may be illustrated best by the following experiment performed with this motor. The motor was run empty, and a load of about 200 h. p.., far exceeding the normal load, was thrown on suddenly. Both armatures of the motor and generator were seen to stop for an instant, the belts slipping over the pulleys, whereupon both came up to the normal speed with the full load, not having been thrown out of synchronism. The experiment could be repeated any number of times. In some cases, the driving power being sufficient, I have been enabled to throw on a load exceeding 8 to 9 times that which the motor was designed to carry, without affecting the speed in the least.

This will be easily understood from the manner in which the current regulation is effected. Assuming the motor to be running without any load, the poles of the armature and field have a certain relative position which is that of the highest self-induction or counter electromotive force. If load be thrown on, the poles are made to recede; the self-induction or counter electromotive force is thereby diminished and more current passed trough the stationary or movable armature-coils. This regulation is very different from that of a direct current motor. In the latter the current is varied by the motor losing a certain number of revolutions in proportion to the load, and the regulation would be impossible if the speed would be maintained constant; here the whole regulation is practically effected during a fraction of one revolution only. From this it is also apparent that it is a practical impossibility to throw such a motor out of synchronism, as the whole work must be done in an instant, it being evident that if the load is not sufficient to make a motor lose a fraction of the first revolution it will not be able to do so in the succeeding revolutions. As to the efficiency of these motors, it is perfectly practicable to obtain 94 to 95 per cent.

The results above given were obtained on a three-wire system. The same motor has been started and operated on two wires in a variety of ways, and although it was not capable of performing quite as much work as on three wires, up to about 60 h. p. it gave results practically the same as those above-mentioned. In fairness to the electricians of Ganz & Co., I must state here that the speed of this motor was higher than that used in their experiments, it being about 1,500. I cannot make due allowance for this difference, as the diameter of the armature and other particulars of the Ganz & Co. motor were not given.

The motor tested had a weight of about 5,000 lbs. From this it will be seen that the performance even on two wires was quite equal to that of the best direct current motors. The motor being of a synchronous type, it might be implied that it was not capable of starting. On the contrary, however, it had a considerable torque on the start and was capable of starting under fair load.

In the article above referred to the assertion is made that the weight of such alternate current motor, for a given capacity, is "several times" larger than that of a direct current motor. In answer to this I will state here that we have motors which with a weight of about 850 pounds develop 10 h. p. with an efficiency of very nearly 90 per cent, and the spectacle of a direct current motor weighing, say 200 – 300 pounds and performing the same work, would be very gratifying for me to behold. The motor which I have just mentioned had no commutator or brushes of any kind nor did it require any direct current.

Finally, in order to refute various assertions made at random, principally in the foreign papers, I will take the liberty of calling to the attention of the critics the fact that since the discovery of the principle several types of motors have been perfected and of entirely different characteristics, each suited for a special kind of work, so that while one may be preferable on account of its ideal simplicity, another might be more efficient. It is evidently impossible to unite all imaginable advantages in one form, and it is equally unfair and unreasonable to judge all different forms according to a common standard. Which form of the existing motors is best, time will show; but even in the present state of the art we are enabled to satisfy any possible demand in practice.

Nikola Tesla
Pittsburgh, Pa.

The Losses Due to Hysteresis in Transformers

The Electrical Engineer - N. Y. — April 9, 1890

In your issue of April 2, in referring to certain remarks made by me at the recent meeting of the American Institute of Electrical Engineers on the subject of hysteresis you make the statement: "It is this constancy of relation that, as Mr. Tesla pointed out may ultimately establish the correctness of the hypothesis advanced, that in reality there is no loss due to hysteresis, and that the changes of magnetization represent a charging and discharging of molecular energy without entailing an actual expenditure of energy."

I do not recollect having made such a statement, and as I was evidently misunderstood, you will greatly oblige me in inserting the following few lines, which express the idea I meant to advance:

Up to the present no satisfactory explanation of the causes of hysteresis has been given. In the most exhaustive and competent treatise on the theory of transformers, by Fleming, static hysteresis is explained by supposing that "the magnetic molecules or molecular magnets, the arrangement of which constitutes magnetization, move stiffly, and the dissipation of energy is the work done in making the necessary magnetic displacement against a sort of magnetic friction." Commonly it is stated that this is a distinct element in the loss of energy in an iron core undergoing magnetic changes entirely independent of any currents generated therein.

Now it is difficult to reconcile these views with our present notions on the constitution of matter generally. The molecules or iron cannot be connected together by anything but elastic forces, since they are separated by an intervening elastic medium through which the forces act; and this being the case is it not reasonable to assume that if a given amount of energy is taken up to bring the molecules out of their original position an equivalent amount of energy should be restored by the molecules reassuming their original positions, as we know is the case in all molecular displacements? We cannot imagine that an appreciable amount of energy should be wasted by the elastically connected molecules swinging back and forth from their original positions, which they must constantly tend to assume, at least within the limit of elasticity, which in all probability is rarely surpassed. The losses cannot be attributed to mere displacement, as this would necessitate the supposition that the molecules are connected rigidly, which is quite unthinkable.

A current cannot act upon the particles unless it acts upon currents in the same, either previously existing or set up by it, and since the particles are held together by elastic forces the losses must be ascribed wholly to the current generated. The remarkable discovery of Ewing that the magnetization is greater on the descent than on the ascent for the same values of magnetizing force strongly points to the fact that hysteresis is intimately connected with the generation of currents either in the molecules individually or in groups of them through the space intervening. The fact observed accords perfectly with our experience on current induction, for we know that on the descent any current set up must be of the same direction with the inducing current, and, therefore, must join with the same in producing a common effect; whereas, on the ascent the contrary is the case.

Dr. Duncan stated that the ratio of increase of primary and secondary current is constant. This statement is, perhaps, not sufficiently expressive, for not only is the ratio constant but, obviously, the differential effect of primary and secondary is constant. Now any current generated - molecular or Foucault currents in the mass - must be in amount proportionate to the difference of the inductive effect of the primary and secondary, since both currents add algebraically - the ratio of windings duly considered, - and as this difference is constant the loss, if wholly accounted for in this manner, must be constant. Obviously I mean here the transformers under consideration, that is, those with a closed magnetic circuit, and I venture to say that the above will be more pronounced when the primary and secondary are wound one on top of the other than when they are wound side by side; and generally it will be the more pronounced the closer their inductive relation.

Dr. Duncan's figures also show that the loss is proportionate to the square of the electromotive force. Again this ought to be so, for an increased electromotive force causes a proportionately increased current which, in accordance with the above statements, must entail a loss in the proportion of the square.

Certainly, to account for all the phenomena of hysteresis, effects of mechanical vibration, the behavior of steel and nickel alloy, etc., a number of suppositions must be made; but can it not be assumed that, for instance, in the case of steel and nickel alloys the dissipation of energy is modified by the modified resistance; and to explain the apparent inconsistency of this view we only need to remember that the resistance of a body as a whole is not a measure of the degree of conductivity of the particles of which it is composed.

N. Tesla
New York City

Swinburne's "Hedgehog" Transformer

The Electrical Engineer - N. Y. — Sept. 24, 1890

Some time ago Mr. Swinburne advanced certain views on transformers which have elicited some comment. In *The Electrical Engineer* of Sept. 10. there are brought out further arguments on behalf of his open circuit, or, as he calls it, "hedgehog" transformer, claiming for this type a higher average efficiency than is attainable with the closed circuit forms. In regard to this, I say with Goethe, *Die Botschaft hör' ich wohl, allein mir fehlt der Glaube* — I hear the message, but I lack belief."

Many of Mr. Swinburne's arguments are in my opinion erroneous. He says: "In calculating the efficiencies of transformers, the loss in the iron has generally been left completely out of account, and the loss in copper alone considered; hence, the efficiencies of 97 and 98 per cent. claimed for closed iron circuit forms." This is a statement little complimentary to those who have made such estimates, and perhaps Mr. Swinburne would be very much embarrassed to cite names on behalf of his argument. He assumes the loss in the iron in the closed circuit forms to be 10 per cent. of the full load, and further "that in most stations the average use of lamps is less than two hours a day, including all lamps installed," and arrives at some interesting figures in regard to efficiency. Mr. Swinburne seems not to be aware of the improvements made in the iron. The loss with the best quality of iron will, I believe, not reach 8 per cent. of the full load by an intelligent use of the transformer, and there is no doubt that further improvements will be made in that direction.

As regards the second part of his assumption, I think that it is exaggerated. It must be remembered that in most central stations or large plants due care is taken that the load is favorably distributed and in many cases the wiring is such that entire circuits may be shut off at certain hours so that there is during these hours no loss whatever in the transformers.'

In his "hedgehog" form of transformer Mr. Swinburne reduces the iron considerably and comes to the conclusion that even in small transformers the iron loss is under one per cent. of the full load, while in the closed circuit forms, it is, according to him, 10 per cent. It would strengthen this argument if the iron would be dispensed with altogether. Mr. Swinburne does not appreciate fully the disadvantages which the open circuit form, operated at the usual period, entails. In order that the loss in tire iron should be reduced to one-tenth, it is necessary to reduce the weight of the iron core to one-tenth and subject every unit length of the same to the same magnoto-motive force. If a higher magneto-motive force is used the loss in the core will — within certain limits, at least — be proportionate to the square of tire magneto-motive force. The remark of Mr. Swinburne, "If the iron circuit is opened, the aides of the embracing core can be removed, so the loss by hysteresis is divided by three," is therefore not true; the loss will be divided by $\frac{F_1^2}{F^2}$ where $F_1 > F$. If the iron of the open circuit form is made up in a closed ring the advantage will be at once apparent, for, since the magnetic resistance will be much reduced, the magneto-motive force required will be correspondingly smaller. It is probable that, say, four Swinburne transformers may be joined in such a way as to form a closed magnetic circuit. In this case the amount of iron and copper would remain the same, but an advantage will be gained as as the total magnetic resistance will be diminished. The four transformers will now demand less excitation and since — under otherwise equal conditions — the gain depends on the square of the existing current, it in by no means insignificant. From the above comparison it is evident that the core of such open circuit transformer should be very short, by far shorter than it appears from the cut in *The Electrical Engineer*.

Mr. Swinburne is in error as to the motives which have caused the tendency to shorten the magnetic circuit in closed circuit transformers. It was principally on account of practical considerations and not to reduce the magnetic resistance, which has little to do with efficiency. If a ring be made of, say, 10 centimeters mean length and 10 square centimeters cross section, .and if it be wound all over with the primary and secondary wires, it will be found that it will give the best result with a certain number of alternations. If, now, a ring is made of the same quality of iron but having, say, 20 centimeters mean length and 10 square centimeters section it will give again the best result with the same number of alternations, and the efficiency will be the same as before, provided that the ring is wound all over with the primary and secondary wires. The space inside of the ring will, in the second case, be increased in proportion to the square of the diameter, and there will be no difficulty in winding on it all the wire required. So the length may be indefinitely increased and a transformer of any capacity made, as long as the ring is completely covered by the primary and secondary wires.

If the wires be wound side by side the ring of smaller diameter will give a better result, and the same will be the case if a certain fraction of the ring is not covered by the wires. It then becomes important to shorten the magnetic circuit. But, since in practice it is necessary to enclose the transformer in a casting, if such a ring be trade, it would have to be protected with a layer of laminated iron, which would increase the cost and loss. It may be inclosed in jars of some, insulating material, as Mr. Swinburne does, belt this is less practicable.

Owing to this, the constructors of the most practical forms, such, for instance, as the Westinghouse transformer, to which the Swinburne reasoning applies, have been prompted to enclose the wires as much as possible with the laminated iron, and then it became important to shorted the magnetic circuit, because in this form only a part of the magnetic circuit is surrounded by the wires, as well as for other practical considerations.

In practice it is desirable to get along with the least length of copper conductor on account of cheapness and regulation. Mr. Swinburne states that in his transformer the loss in iron is under one per cent. of the full load; all the balance of loss must, therefore, be in the copper. But since, according to him, the wires are of larger section, his transformer can hardly be an improvement in that direction. The gun-metal casting is also objectionable. There is no doubt some loss going on in the came, and beside, it increases the resistance of the wires by a factor $\frac{S}{S_1}$ where S is the total cross-section of the core and S_1 the section of the iron wires. There is one important point which seems to have escaped Mr. Swinburne's attention. Whether the open circuit transformer is an improvement, or not, will depend principally on the period. The experience of most electrical engineers has resulted in the adoption of the closed-circuit transformer. I believe that I was the first to advocate in open circuit form, but to improve its efficiency I had to use a much higher period; at usual periods the, closed circuit form is preferable.

Mr. Swinburne makes some other obscure statements upon which I need not dwell, as they have no bearing on the main question.

Tesla's New Alternating Motors
The Electrical Engineer - N. Y. — Sept. 24, 1890

I hope you will allow me the privilege, to say in the columns of your esteemed journal a few words in regard to an article which appeared in *Industries* of August 22, to which my attention has been called. In this article an attempt is made to criticize some of my inventions, notably those which you have described in your issue of August 6, 1890.

The writer begins by stating: "The motor depends on a shifting of the poles under certain conditions, a principle which has been *already* employed by Mr. A. Wright in his alternating current meter." This is no surprise to me. It would rather have surprised me to learn that Mr. Wright has *not yet* employed the principle in his meter, considering what, before its appearance, was known of my work on motors, and more particularly of that of Schallenberger on meters. It has cost me years of thought to arrive at certain results, by many believed to be unattainable), for which there are now numerous claimants, and the number of these is rapidly increasing, like that of the colonels in the South after the war.

The writer then good-naturel explains the theory of action of the motive device in Wright's meter, which has greatly benefited me, for it is so long since I have arrived at this, and similar theories, that I had almost forgotten it. He then says: " Mr. Tesla has worked out some more or less complicated motors on this principle, but the curious point is that he has completely misunderstood the theory of the phenomena, and has got hold of (the old fallacy of screening." This may be *curious,* but how much *more curious* it is to find that the writer in *Industries* has *completely misunderstood* everything himself. I like nothing better than just criticism of my work, even if it be severe, but when the critic assumes a certain " l'état c'est moi" air of unquestioned Competency I want him to know what he is writing about. How little the writer in *Industries* seems to know about the matter is painfully apparent when ho connects the phenomenon in Wright's meter with the subject he has under consideration. His further remark, "He (Mr. Tesla) winds his secondary of iron instead of copper and thinks the effect is produced magnetically," is illustrative of the care with which he has perused the description of the devices contained in the issue of *The Electrical Engineer* above referred to.

I take a motor having, say eight poles, and wrap the exciting coils of four alternate cores with fine insulated iron wire. When the current is started in these coils it encounters the effect of the closed magnetic circuit and is retarded. The magnetic lines set up at the start close to the iron wire around the coils and no free poles appear at first at the ends of the four cores. As the current rises in the coils more lines are set up, which crowd more and more in the fine iron wire until finally the same becomes saturated, or nearly so, when the shielding action of the iron wire ceases and free poles appear at the ends of the four protected cores. The effect of the iron wire, as will be seen, is two-fold. First, it retards the energizing current; and second, it delays the appearance of the free poles. To produce still greater difference of phase in the magnetization of the protected and unprotected cores, I connect the iron wire surrounding the coils of the former in series with the coils of the latter, in which case, of course, the iron wire is preferably wound or connected differentially, after the fashion of the resistance, coils in a bridge, so as to have no appreciable self-induction. In other cases I obtain the desired retardation in the appearance of the free poles on one set of cores by *a* magnetic shunt, which produces a greater retardation of the current and takes up at the start a certain number of the lines set up, but becomes saturated when the current in the exciting coils reaches a predetermined strength.

In the transformer the same principle of shielding is utilized. A primary conductor is surrounded with a fine layer of laminated iron, consisting of fine iron wire or plates properly insulated and interrupted. As long as the current in the primary conductor is so small that the iron enclosure can carry all the lines of force set up by the current, there is very little action exerted upon a secondary conductor placed in vicinity to the first; but just as soon as the iron enclosure becomes saturated, or nearly so, it loses the virtue of protecting the secondary and the inducing action of the primary practically begins. What, may I ask, has all this to do with the "old fallacy of screening?"

With certain objects in view — the enumeration of which would lead me too far — an arrangement was shown in *The Electrical Engineer*, about which the writer in *Industries* says : " A ring of laminated iron is wound with a secondary. It is then encased in iron laminated in the *wrong direction* and the primary is wound outside of this. The layer of iron between the primary and secondary is supposed to screen the coil. Of course it cannot do so, such a thing Is unthinkable." This reminds me of the man who had committed some offense and engaged the services of an attorney. "They cannot commit you to prison for that," said the attorney. Finally the man *was* imprisoned. He sent for the attorney. " Sir," said the latter, " I tell you they *cannot* imprison you for that." "But, sir," retorted the prisoner, "they *have* imprisoned me." It *may not* screen, in the opinion of the writer in *Industries,* but just the same it *does.* According to the arrangement the *principal* effect of the screen may be either a retardation of the action of the primary current upon the secondary circuit or a deformation of the secondary current wave with similar results for the purposes intended. In the arrangement referred to by the writer in *Industries* be seems to be certain that the iron layer acts like a choking coil; there again he is mistaken; it does not act like a choking coil, for then its capacity for maintaining constant current would be very limited. But it acts more like a magnetic shunt in constant current transformers and dynamos, as, in my opinion, it ought to act.

There are a good many more things to be said about the remarks contained in *Industries.* In regard to the magnetic time lag the writer says: " If a bar of iron has a coil at one end, and if the core is perfectly laminated, on starting a current in the coil the induction *all along the iron* corresponds to the excitation at that instant, unless there is a microscopic time lag, *of which there is no evidence.*"Yet a motor was described, the very operation of which is dependent on the time lag of magnetization of the different parts of a core. It is true the writer uses the term "perfectly laminated" (which, by the way, I would like him to explain), but if he intends to make such a "perfectly laminated" core I venture to say there is trouble in store for him. From his remarks I see that the writer completely overlooks the importance of the size of the core and of the number of the alternations pointed out; be fails to see the stress laid on the saturation of the screen, or shunt, in some of the cases described; he does not seem to recognize the fact that in the cases considered the formation of current is reduced as far as practicable in the screen, and that the same, therefore, so far as its quality of screening is concerned, has no role to perform as *a conductor.* I also see that he would want considerable information about the time lag in the magnetization of the different parts of a core, and an explanation why, in the transformer he refers to, the screen is laminated in the *wrong direction,* etc. — but the elucidation of all these points would require more time than I am able to devote to the subject. It is distressing to find all this in the columns of a leading technical journal.

In conclusion, the writer shows his true colors by making the following withering remarks: "It is questionable whether the Tesla motor will ever be a success. Such motors will go round, of course, and will give outputs, but their efficiency is doubtful; and if they need three-wire circuits and special generators there is no object in using them, as a direct current motor can be run instead with advantage."

No man of broad views will feel certain of the success of any invention, however good and original, in this period of feverish activity, when every day may bring new and unforseen developments. At the pace we are progressing the permanence of all our apparatus it its present forms becomes more and more problematical. It is impossible to foretell what type of motor will crystalize out of the united efforts of many able men; but it is my conviction that at no distant time a motor having commutator and brushes will be looked upon as an antiquated piece of mechanism. Just how much the last quoted remarks of the writer of *Industries* — considering the present state of the art — are justified, I will endeavor to show in a few lines.

First, take the transmission of power in isolated places. A case frequently occurring in practice and attracting more and more the attention of engineers is the transmission of large powers at considerable distances. In such a case the power is very likely to be cheap, and the cardinal requirements are then the reduction of the cost of the leads, cheapness of construction and maintenance of machinery and constant speed of the motors. Suppose a loss of only 25 per cent in the leads, at full load, be allowed. If a direct current motor be used, there will be, besides other difficulties, considerable variation in the speed of the motor — even if the current is supplied from a series dynamo — so much so that the motor may not be well adapted for many purposes, for instance, in cases where direct current transformation is contemplated with the object of running lights or other devices at constant potential. It is true that the condition may be bettered by employing proper regulating devices, but these will only further complicate the already complex system, and in all probability fail to secure such perfection as will be desired. In using an ordinary single-circuit alternate current motor the disadvantage is that the motor has no starting torque and that, for equal weight, its output and efficiency are more or less below that of a direct current motor. If, on the contrary, the armature of any alternator or direct current machine — large, low-speed, two-pole machines will give the best results — is wound with two circuits, a motor is at once obtained which possesses sufficient torque to start under considerable load: it runs in absolute synchronism with the generator — an advantage much desired and hardly ever to be attained with regulating devices; it takes current in proportion to the load, and its plant efficiency within a few per cent is equal to that of a direct current motor of the same size. It will be able, however, to perform more work than a direct current motor of the same size, first, because there will be no change of speed, even if the load be doubled or tripled, within the limits of available generator power; and second, because it can be run at a higher electromotive force, the commutator and the complication and difficulties it involves in the construction and operation of the generators and motors being eliminated from the system. Such a system will, of course, require three leads, but since the plant efficiency is practically equal to that of the direct current system, it will require the same amount of copper which would be required in the latter system, and the disadvantage of the third lead will be comparatively small, if any, for three leads of smaller size may perhaps be more convenient to place than two larger leads. When more machines have to be used there may be no disadvantage whatever connected with the third wire ; however, since the simplicity of the generators and motors allows the use of higher electromotive forces, the cost of the leads may be reduced below the figure practicable with the direct current system.

Considering all the practical advantages offered by such an alternating system, I am of an opinion quite contrary to that of the author of the article in *Industries,* and think that it can quite successfully stand the competition of any direct current system, and this the more, the larger the machines built and the greater the distances.

Another case frequently occurring in practice is the transmission of small powers in numerous isolated places, such as mines, etc. In many of these cases simplicity and reliability of the apparatus are the principal objects. I believe that in many places of this kind my motor has so far proved a perfect success. In such cases a type of motor is used possessing great starting torque, requiring for its operation only alternating current and having no sliding contacts whatever on the armature, this advantage over other types of motors being highly valued in such places. The plant efficiency of this form of motor is, in the present state of perfection, inferior to that of the former form, but I am confident that improvements will be made in that direction. Be-sides, plant efficiency is in these cases of secondary importance, and in cases of transmission at considerable distances, it is no drawback, since the electromotive force may be raised as high as practicable on converters. I can not lay enough stress on this advantageous feature of my motors, and should think that it ought to be fully appreciated by engineers, for to high electromotive forces we are surely coming, and if they must be used, then the fittest apparatus will be employed. I believe that in the transmission of power with such commutatorless machines, 10,000 volts, and even more, may be used, and I would be glad to see Mr. Ferranti's enterprise succeed. His work is in the right direction, and, in my opinion, it will be of great value for the advancement of the art.

As regards the supply of power from large central stations in cities or centers of manufacture, the above arguments are applicable, and I see no reason why the three-wire motor system should not be successful. In putting up such a station, the third wire would be but a very slight drawback, and the system possesses enough advantages to over-balance this and any other disadvantage. But this question will be settled in the future, for as yet comparatively little has been done in that direction, even with the direct current system. The plant efficiency of such a three-wire system would be increased by using, in connection with the ordinary type of my motor, other types which act more like inert resistances. The plant efficiency of the whole system would, in all cases, be greater than that of each individual motor — if like motors are used — owing to the fact that they would possess different self-induction, according to the load.

The supply of power from lighting mains is, I believe, in the opinion of most engineers, limited to comparatively small powers, for obvious reasons. As the present systems are built on the two-wire plan, an efficient two-wire motor without commutator is required for this purpose, and also for traction purposes. A large

number of these motors, embodying new principles, have been devised by me and are being constantly perfected. On lighting stations, however, my three-wire system may be advantageously carried out. A third wire may be run for motors and the old connections left undisturbed. The armatures of the generators may be rewound, whereby the output of the machines will be increased about 35 per cent, or even more in machines with cast iron field magnets. If the machines are worked at the same capacity, this means an increased efficiency. If power is available at the station, the gain in current may be used in motors. Those who object to the third wire, may remember that the old two-wire direct system is almost entirely superseded by the three-wire system, yet my three-wire system offers to the alternating system relatively greater advantages, than the three-wires direct possesses over the two-wire. Perhaps, if the writer in *Industries* would have taken all this in consideration, he would have been less hasty in his conclusions.
Nikola Tesla
New York
Sept. 17, 1890

Experiments with Alternating Currents of High Frequency

The Electrical Engineer - N. Y. — March. 18, 1891

In *The Electrical Engineer* issue of 11th inst., I find a note of Prof. Elihu Thomson relating to some of my experiments with alternating currents of very high frequency.

Prof. Thomson calls the attention of your readers to the interesting fact that he has performed some experiments in the same line. I was not quite unprepared to hear this, as a letter from him has appeared in the *Electrician* a few months ago, in which he mentions a small alternate current machine which was capable of giving, I believe, 5,000 alternations per second, from which letter it likewise appears that his investigations on that subject are of a more recent date.

Prof. Thomson describes an experiment with a bulb enclosing a carbon filament which was brought to incandescence by the bombardment of the molecules of the residual gas when the bulb was immersed in water. "rendered slightly conducting by salt dissolved therein," (?) and a potential of 1,000 volts alternating 5,000 time a second applied to the carbon strip. Similar experiments have, of course, been performed by many experimenters, the only distinctive feature in Prof. Thomson's experiment being the comparatively high rate of alternation. These experiments can also be performed with a steady difference of potential between the water and the carbon strip in which case, of course, conduction through the glass takes place, the difference of potential required being in proportion to the thickness of the glass. With 5,000 alternations per second, conduction still takes place, but the condenser effect is preponderating. It goes, of course, without saying that the healing of the glass in such a case is principally due to the bombardment of the molecules, partly also to leakage or conduction, but it is an undeniable fact that the glass may also be heated merely by the molecular displacement. The interesting feature in ray experiments was that a lamp would light up when brought near to an induction coil, and that it could be held in the hand and the filament brought to incandescence.

Experiments of the kind described I have followed up for a long time with some practical objects in view. In connection with the experiment described by Prof. Thomson, if may be of interest to mention a very pretty phenomenon which may be observed with an incandescent lamp. If a lamp be immersed in water as far as practicable and the filament and the vessel connected to the terminals of an induction coil operated from a machine such as I have used in my experiments, one may see the dull red filament surrounded by a very luminous globe around which there is a less luminous space. The effect is probably due to reflection, as the globe is sharply defined, but may also be due to a "dark space;" at any rate it is so pretty that it must be seen to be appreciated.

Prof. Thomson has misunderstood my statement about the limit of audition. I was perfectly well aware of the fact that opinions differ widely on this point. Nor was I surprised to find that arcs of about 10,000 impulses per second, emit a sound. My statement " the *curious* point is," etc. was only made in deference to an opinion expressed by Sir William Thomson. There was absolutely no stress laid on the precise number. The popular belief was that something like 10,000 to 20,000 per second, or 20,000 to 40,000, at the utmost was the limit. For my argument this was immaterial. I contended that sounds of an incomparably greater number, that is, many times even the highest number, could be heard if they could be produced with sufficient power. My statement was only speculative, but I have devised means which I think may allow me to learn something definite on that point. I have not the least doubt that it is simply a question of power. A very short arc may be silent with 10,000 per second, but just as soon as it is lengthened it begins to emit a sound. The vibrations are the same in number, but more powerful.

Prof. Thomson states that I am taking as the limit of " audition sounds from 5,000 to 10,000 complete waves per second." There is nothing in my statements from which the above could be inferred, but Prof. Thomson has perhaps not thought that there are two sound vibrations for each complete current wave, the former being independent of the direction of the current.

I am glad to learn that Prof. Thomson agrees with me as to the causes of the persistence of the arc. Theoretical considerations considerable time since have led me to the belief that arcs produced by currents of such high frequency would possess this and other desirable features. One of my objects in this direction has been to produce a practicable small arc. With these, currents, for many reasons, much smaller arcs are practicable.

The interpretation by Prof. Thomson of my statements about the arc system leads me *now*, he will pardon me for saying so, to believe that what is most essential to the success of an arc system is a good management. Nevertheless I feel confident of the correctness of the views expressed. The conditions in practice are so manifold that it is impossible for any type of machine to prove best in all the different conditions.

In one case, where the circuit is many miles long, it is desirable to employ the most efficient machine with the least internal resistance ; in another case such a machine would not be the best to employ. It will certainly be admitted that a machine of any type must have a greater resistance if intended to operate arc lights than if it is designed to supply incandescent lamps in series. When arc lights are operated and the resistance is small, the lamps are unsteady, unless a type of lamp is employed in which the carbons are separated by a mechanism which has no further influence upon the feed, the feeding being effected by an independent mechanism ; but even in this case the resistance must be considerably greater to allow a quiet working of the lamps. Now, if the machine be such as to yield a steady current, there is no way of attaining the desired result except by putting the required resistance somewhere either inside or outside of the machine. The latter is hardly practicable, for the customer may stand a hot machine, but he looks with suspicion upon a hot resistance box. A good automatic regulator of course improves the machine and allows us to reduce the internal resistance to some extent, but not as far as would be desirable. Now, since resistance is loss, we can advantageously replace resistance in the machine by an equivalent impedance. But to produce a great impedance with small ohmic resistance, it is necessary to have self-induction and variation of current, and the greater the self-induction and the rate of change of the current, the greater the impedance may be made, while the ohmic resistance may be very small. It may also be remarked that the impedance of the circuit external to the machine is likewise increased. As regards the increase in ohmic resistance in consequence of the variation of the current, the same is, in the commercial machines now in use, very small. Clearly then a great advantage is gained by providing self induction in the machine circuit and undulating the current, for it is possible to replace a machine which has a resistance of, say, 16 ohms by one which has no more than 2 or 3 ohms, and the lights will work even steadier. It seems to me therefore, that my saying that self-induction is essential to the commercial success of an arc system is justified. What is still more important, such a machine will cost considerably less. But to realize fully the benefits, it is preferably to employ an alternate current machine, as in this case a greater rate of change in the current is obtainable. Just what the ratio of resistance to impedance is in the Brush and Thomson machines is nowhere stated, but I think that it is smaller in the Brush machine, judging from its construction.

As regards the better working of clutch lamps with undulating currents, there is, according to my experience, not the least doubt about it. I have proved it on a variety of lamps to the complete satisfaction not only of myself, but of many others. To see the improvement in the feed and to the jar of the clutch at its best it is desirable to employ a lamp in which an independent clutch mechanism effects the feed, and the release of the rod is independent of the up and down movement. In such a lamp the clutch has a small inertia and is very sensitive to vibration, whereas, if the feed is effected by the up and down movement of the lever carrying the rod, the inertia of the system is so great that it is not affected as much by vibration, especially if, as in many cases, a dash pot is employed. During the year 1885 I perfected such a lamp which wan calculated to be operated with undulating currents. With about 1,500 to 1,800 current impulses per minute the feed of this lamp is such that absolutely no movement of the rod can be observed, even if the arc be magnified fifty-fold by means of a lens; whereas, if a steady current is employed, the lamp feeds by small steps. I have, however, demonstrated this feature on other types of lamps, among them being a derived circuit lamp such as Prof. Thomson refers to. I conceived the idea of such a lamp early in 1884, and when my first company was started, this was the first lamp I perfected. It was not until the lamp was ready for manufacture that, on receiving copies of applications from the Patent Office, I learned for the first time, not having had any knowledge of the state of the art in America, that Prof. Thomson had anticipated me and had obtained many patents on this principle, which, of course, greatly disappointed and embarrassed me at that time. I observed the improvement of the feed with undulating currents on that lamp, but I recognized the advantage of providing a light and independent clutch unhampered in its movements. Circumstances did not allow me to carry out at that time some designs of machines I had in mind, and with the existing machines the lamp has worked at a great disadvantage. I cannot agree with Prof. Thomson that small vibrations would benefit a clockwork lamp as much as a clutch lamp; in fact, I think that they do not at all benefit a clockwork lamp.

It would be interesting to learn the opinion of Mr. Charles F. Brush on these points.

Prof. Thomson states that he has run with perfect success clutch lamps " in circuit with coils of such large self-induction that any but very slight fluctuations were wiped out." Surely Prof. Thomson does not mean to say that self-induction wiped out the periodical fluctuations of the current. For this, just the opposite quality, namely, capacity, is required. The self-induction of the coils in this case simply augmented the impedence and prevented the great variations occurring at large time intervals, which take place when the resistance in circuit with the lamps is too small, or even with larger resistance in circuit when the dash pots either in the lamps or elsewhere are too loose.

Prof. Thomson further states that in a lamp in which the feed mechanism is under the control of the derived circuit magnet only, the fluctuations pass through the arc without affecting the magnet to a perceptible degree. It is true that the variations of the resistance of the arc, in consequence of the variations in the current strength, are such as to dampen the fluctuation. Nevertheless, the periodical fluctuations are transmitted through the derived circuit, as one may convince himself easily of, by holding a thin plate of iron against the magnet.

In regard to the physiological effects of the currents I may state that upon reading the memorable lecture of Sir William Thomson, in which he advanced his views on the propagation of the alternate currents through conductors, it instantly occurred to me that currents of high frequencies would be less injurious. I have been looking for a proof that the mode of distribution through the body is the cause of the smaller physiological effects. At times I have thought to have been able to locate the pain in the outer portions of the body, but it is very uncertain. It is most certain, however, that the feeling with currents of very high frequencies is somewhat different from that with low frequencies. I have also noted the enormous importance of one being prepared for the shock or not.

If one is prepared, the effect upon the nerves is not nearly as great as when unprepared. With alternations as high as 10,000 per second and upwards, one feels but little pain in the central portion of the body. A remarkable feature of such currents of high tension is that one receives a burn instantly he touches the wire, but beyond that the pain is hardly noticeable.

But since the potential difference across the body by a given current through it is very small, the effects can not be very well ascribed to the surface distribution of the current, and the excessively low resistance of the body to such rapidly varying currents speak rather for a condenser action.

In regard to the suggestion of Dr. Tatum, which Prof. Thomson mentions in another article in the same issue, I would state that I have constructed machines up to 480 poles, from which it is possible to obtain about 30,000 alternations per second, and perhaps more. I have also designed types of machines in which the field would revolve in an opposite direction to the armature, by which means it would be possible to obtain from a similar machine 60,000 alternations per second or more.

I highly value the appreciation of Prof. Thomson of my work, but I must confess that in his conclusion he makes a most astounding statement as to the motives of his critical remarks. I have never for a moment thought that his remarks would be dictated by anything but friendly motives. Often we are forced in daily life to represent opposing interests or opinions, but surely in the higher aims the feelings of friendship and mutual consideration should not be affected by such things as these.

N. Tesla

Alternate Current Motors

The Electrical Engineer - April 3, 1891

Sir — In your issue of March 6 I find the passage: "Mr. Kapp described the position as it exists. He showed how Ferraris first of all pointed out the right way to get an alternating-current motor that was self-starting, and how Tesla and others had worked in the direction indicated by Ferraris," etc.

I would be very glad to learn how Mr. Kapp succeeded in showing this. I may call his attention to the fact that the date of filing of my American patent anticipates the publication of the results of Prof. Ferraris in Italy by something like six months. The date of filing of my application is, therefore, the first public record of the invention. Considering this fact, it seems to me that it would be desireable that Mr. Kapp should modify his statement. —Yours, etc.

Nikola Tesla

New York, 17th March , 1891

Electro-Motors

Electrical Review - London — April 3, 1891

Fifteen or sixteen years ago, when I was pursuing my course at the college, I was told by an eminent physicist that a motor could not he operated without the use of brushes and commutators, or mechanical means of some kind for commutating the current. It was then I determined to solve the problem.

After years of persistent thought I finally arrived at a solution. I worked out the theory to the last detail, and confirmed all of my theoretical conclusions by experiments. Recognizing the value of the invention, I applied myself to the work of perfecting it, and after long continued labor I produced several types of practical motors.

Now all this I did long before anything whatever transpired in the whole scientific literature — as far as it could be ascertained — which would have even pointed at the possibility of obtaining such a result, but quite contrary at a time when scientific and practical men alike considered this result unattainable. In all civilized countries patents have been obtained almost without a single reference to anything which would have in the least degree rendered questionable the novelty of the invention. The first published essay — an account of some laboratory experiments by Prof. Ferraris — was published in Italy six or seven months after the date of filing of my applications for the foundation patents. The date of filing of my patents is thus the first public record of the invention. Yet in your issue of March 6th I read the passage: " For several years past, from the days of Prof. Ferraris's investigations, which were followed by those of Tesla, Zipernowsky and a host of imitators," etc.

No one can say that I have not been free in acknowledging the merit of Prof. Ferraris, and I hope that my statement of facts will not be misinterpreted. Even if Prof. Ferraris's essay would have anticipated the date of filing of my application, yet, in the opinion of all fairminded men, I would have been entitled to the credit of having been the first to produce a practical motor ; for Prof. Ferraris himself denies in his essay the value of the invention for the transmission of power, and only points out the possibility of using a properly-constructed generator, which is the only way of obtaining the required difference of phase without losses; for even with condensers — by means of which it is possible to obtain a quarter phase — there are considerable losses, the cost of the condensers not considered.

Thus, in the most essential features of the system — the generators with the two or three circuits of differing phase, the three-wire system, the closed coil armature, the motors with direct current in the field, etc. — I would stand alone, even had Prof. Ferraris's essay been published many years ago.

As regards the most practicable form of two-wire motor, namely, one with a single energising circuit and induced circuits, of which there are now thousands in use, I likewise stand alone.

Most of these facts, if not all, are perfectly well known in England; yet, according to some papers, one of the leading English electricians does not hesitate to say that I have worked in the direction indicated by Prof. Ferraris, and in your issue above referred to it seems I am called an imitator.

Now, I ask you where is that world-known English fairness ? I am a pioneer, and I am called an imitator. I am not an imitator. I produce original work or none at all.

Nikola Tesla

Phenomena of Currents of High Frequency

The Electrical Engineer - N. Y. — April. 8, 1891

I cannot pass without comment the note of Prof. Thomson in your issue of April 1, although I dislike very much to engage in a prolonged controversy. I would gladly let Prof. Thomson have the last word, were it not that some of his statements render a reply from me necessary.

I did not mean to imply that, whatever work Prof. Thomson has done in alternating currents of very high frequency, he had done subsequent to his letter published in the *Electrician*. I thought it possible, and even probable, that he had made his experiments some time before, and my statement in regard to this was meant in this general way. It is more than probable that quite a number of experimenters have built such machines and observed effects similar to those described by Prof. Thomson. It is doubtful, however, whether, in the absence of any publication on this subject, the luminous phenomena described by me have been observed by others, the more so, as very few would be likely to go to the trouble I did, and I would myself not have done so had I not had beforehand the firm conviction, gained from the study of the works of the most advanced thinkers, that I would obtain the results sought for. Now, that I have indicated the direction, many will probably follow, and for this very purpose I have shown some of the results I have reached.

Prof. Thomson states decisively in regard to experiments with the incandescent lamp bulb and the filament mounted on a single wire, that he cannot agree with me at all that conduction through the glass has anything to do with the phenomenon observed. He mentions the well-known fact that an incandescent lamp acts as a Leyden jar and says that " if conduction through the glass were a possibility this action could not occur." I think I may confidently assert that very few electricians will share this view. For the possibility of the condenser effect taking place it is only necessary that the rate at which the charges can equalize through the glass by conduction should be somewhat below the rate at which they are stored.

Prof. Thomson seems to think that conduction through the glass is an impossibility. Has he then never measured insulation resistance, and has he then not measured it by means of a conduction current? Does he think that there *is* such a thing as a perfect non-conductor among the bodies we are able to perceive ? Does he *not* think that as regards conductivity there can be question only of degree? If glass were a perfect non-conductor, how could we account for the leakage of a glass condenser when subjected to steady differences of potential?

While not directly connected with the present controversy, I would here point out that there exists a popular error in regard to the properties of dielectric bodies. Many electricians frequently confound the theoretical dielectric of Maxwell with the dielectric bodies in use. They do not stop to think that the only perfect dielectric is ether, and that all other bodies, the existence of which is known to us, must be conductors, judging from their physical properties.

My statement that conduction is concerned to some, although perhaps negligible, extent in the experiment above described was, however, made not only on account of the fact that all bodies conduct more or less, but principally on account of the heating of the glass during the experiment. Prof. Thomson seems to overlook the fact that the insulating power of glass diminishes enormously with the increase in temperature, so much so, that melted glass is comparatively an excellent conductor. I have, moreover, stated in my first reply to Prof. Thomson in your issue of March 18, that the same experiment can be performed by means of an unvarying difference of potential. In this case it must be assumed that some such process as conduction through the glass takes place, and all the more as it is possible to show by experiment, that with a sufficiently high steady difference of potential, enough current can be passed through the glass of a condenser with mercury coating to light up a Geissler tube joined in series with the condenser. When the potential is alternating, the condenser action comes in and conduction becomes insignificant, and the more so, the greater the rate of alternation or change per unit of time. Nevertheless, in my opinion, conduction must always exist, especially if the glass is hot, though it may be negligible with very high frequencies.

Prof. Thomson states, further, that from his point of view I have misunderstood his statement about the limit of audition. He says that 10,000 to 20,000 alternations correspond to 5,000 to 10,000 complete waves of sound. In my first reply to Prof. Thomson's remarks (in your issue of March 18,) I avoided pointing out directly that Prof. Thomson was mistaken, but now I see no way out of it. Prof. Thomson will pardon me if I call his attention to the fact he seems to disregard, namely, that 10,000 to 20,000 alternations of current in an arc — which was the subject under discussion — do not mean 5,000 to 10,000, but 10,000 to 20,000 *complete waves of sound.*

He says that I have adopted or suggested as the limit of audition10,000 waves per second, but I have neither adopted nor suggested it. Prof. Thomson states that I have been working with 5,000 to 10,000 complete waves, while I have nowhere made any such statement. He says that this would be working below the limit of audition, and cites as an argument that at the Central High School, in Philadelphia, he has heard 20,000 waves per second ; but he wholly overlooks a point on which I have dwelt at some length, namely, that the limit of audition of an arc is something entirely different from the limit of audition in general.

Prof. Thomson further states, in reply to some of my views expressed in regard to the constant current machines that five or six years ago it occurred to him to try the construction of a dynamo for constant current, in which " the armature coils were of a highly efficient type, that is, of comparatively short wire length for the voltage and moving in a dense magnetic field." Exteriorly to the coils and to the field he had placed in the circuit of each coil an impedance coil which consisted of an iron core wound with a considerable length of wire and connected directly in circuit with the armature coil. He thus obtained, he thought, " the property of considerable self-induction along with efficient current generation." Prof. Thomson says he expected " that possibly the effects would be very much the same as those obtainable from the regularly constructed apparatus." But he was disappointed, he adds. With all the consideration due to Prof. Thomson, I would say that, to expect a good result from such a combination, was rather sanguine. Earth is not farther from Heaven than this arrangement is from one, in which there would be a length of wire, sufficient to give the same self-induction, wound on the armature and utilized to produce useful E. M. F., instead of doing just the opposite, let alone the loss in the iron cores. But it is, of course, only fair to remember that this experiment was performed five or six years ago, when even the foremost electricians lacked the necessary information in these and other matters.

Prof. Thomson seems to think that self-induction wipes out the periodical undulations of current. Now self-induction does not produce any such effect, but, if anything, it renders the undulation more pronounced. This is self-evident. Let us insert a self-induction coil in a circuit traversed by an undulating current and see what happens. During the period of the greatest rate of change, when the current has a small value, the self-induction opposes more than during the time of the small rate of change when the current is at, or near, its maximum value. The consequence is, that with the same frequency the maximum value of the current becomes the greater, the greater the self-induction. As the sound in a telephone depends only on the maximum value, it is clear that self-induction is the very thing required in a telephone circuit. The larger the self-induction, the louder and clearer the speech, provided the same current is passed through the circuit. I have had ample opportunity to study this subject during my telephone experience of several years. As regard the fact that a self-induction coil in series with a telephone diminishes the loudness of the sound, Prof. Thomson seems to overlook the fact that this effect is wholly due to the impedance of the coil, *i. e.,* to its property of diminishing the *current strength.* But while the current strength is diminished the undulation is rendered only more pronounced. Obviously, when comparisons are made they must be made with the same current.

In an arc machine, such as that of Prof. Thomson's, the effect is different. There, one has to deal with a make and break. There are then two induced currents, one in the opposite, the other in the same direction with the main current. If the function of the mechanism be the same whether a self-induction coil be present or not, the undulations could not possibly be wiped out. But Prof. Thomson seems, likewise, to forget that the effect is wholly due to the defect of the commutator ; namely, the induced current of the break, which is of the same direction with the main current and of great intensity, when large self-induction is present, simply bridges the adjacent commutator segments, or, if not entirely so, at least shortens the interval during which the circuit is open and thus reduces the undulation.

In regard to the improvement in the feeding of the lamps by vibrations or undulations, Prof. Thomson expresses a decisive opinion. He now says that the vibrations *must* improve the feeding of a clock-work lamp. He says that I "contented myself by simply saying," that I cannot agree with him on that point.

Now, sa*ying it,* is not the only thing I did. I have passed many a night watching a lamp feed, and I leave it to any skilled experimenter to investigate whether my statements are correct. My opinion is, that a clock-work lamp; that is, a lamp in which the descent of the carbon is regulated not by a clutch or friction mechanism, but by an escapement, cannot feed any more perfectly than tooth by tooth, which may be a movement of, say, 1/15 of an inch or less. Such a lamp will feed in nearly the same manner whether the current be perfectly smooth or undulating, providing the conditions of the circuit are otherwise stable. If there is any advantage, I think it would be in the use of a smooth current, for, with an undulating current, the lamp is likely to miss some time and feed by more than one tooth. But in a lamp in which the descent of the carbon is regulated by friction mechanism, an indulating current of the proper number of undulations per second will always give a better result. Of course, to realize fully the benefits of the undulating current the release ought to be effected independently of the up-and-down movement I have pointed out before.

In regard to the physiological effects, Prof. Thomson says, that in such a comparatively poor conductive material as animal tissue the distribution of current cannot be governed by self-induction to any appreciable extent, but he does not consider the two-fold effect of the large cross-section, pointed out by Sir William Thomson. As the resistance of the body to such currents is low, we must assume either condenser action or induction of currents in the body.

Nikola Tesla

New York, April 4, 1891

Electric Discharge in Vacuum Tubes

The Electrical Engineer - N.Y. — July 1, 1891

In *The Electrical Engineer* of June 10 I have noted the description of some experiments of Prof. J. J. Thomson, on the "Electric Discharge in Vacuum Tubes," and in your issue of June 24 Prof. Elihu Thomson describes an experiment of the same kind. The fundamental idea in these experiments is to set up an electromotive force in a vacuum tube — preferably devoid of any electrodes — by means of electromagnetic induction, and to excite the tube in this manner.

As I view the subject I should think that to any experimenter who had carefully studied the problem confronting us and who attempted to find a solution of it, this idea must present itself as naturally as, for instance, the idea of replacing the tinfoil coatings of a Leyden jar by rarefied gas and exciting luminosity in the condenser thus obtained by repeatedly charging and discharging it. The idea being obvious, whatever merit there is in this line of investigation must depend upon the completeness of the study of the subject and the correctness of the observations. The following lines are not penned with any desire on my part to put myself on record as one who has performed `similar experiments, but with a desire to assist other experimenters by pointing out certain peculiarities of the phenomena observed, which, to all appearances, have not been noted by Prof. J. J. Thomson, who, however, seems to have gone about systematically in his investigations, and who has been the first to make his results known. These peculiarities noted by me would seem to be at variance with the views of Prof. J. J. Thomson, and present the phenomena in a different light.

My investigations in this line occupied me principally during the winter and spring of the past year. During this time many different experiments were performed, and in my exchanges of ideas on this subject with Mr. Alfred S. Brown, of the Western Union Telegraph Company, various different dispositions were suggested which were carried out by me in practice. Fig. 1 may serve as an example of one of the many forms of apparatus used. This consisted of a large glass tube sealed at one end and projecting into an ordinary incandescent lamp bulb. The primary, usually consisting of a few turns of thick, well-insulated copper sheet was inserted within the tube, the inside space of the bulb furnishing the secondary. This form of apparatus was arrived at after some experimenting, and was used principally with the view of enabling me to place a polished reflecting surface on the inside of the tube, and for this purpose the last turn of the primary was covered with a thin silver sheet. In all forms of apparatus used there was no special difficulty in exciting a luminous circle or cylinder in proximity to the primary. ,

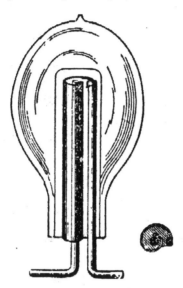

As to the number of turns, I cannot quite understand why Prof. J. J. Thomson should think that a few turns were "quite sufficient," but lest I should impute to him an opinion he may not have, I will add that I have gained this impression from the reading of the published abstracts of his lecture. Clearly, the number of turns which gives the best result in any case, is dependent on the dimensions of the apparatus, and, were it not for various considerations, one turn would always give the best result.

I have found that it is preferable to use. in these experiments an, alternate current machine giving a moderate number of alternations per second to excite the induction coil for charging the Leyden jar which discharges through the primary — shown diagrammatically in Fig. 2, — as in such case, before the disruptive discharge takes place, the tube or bulb is slightly excited and the formation of the luminous circle is decidedly facilitated. But I have also used a Wimshurst machine in some experiments.

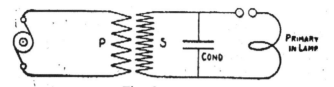

Prof. J. J. Thomson's view of the phenomena under consideration seems to be that they are wholly due to electro-magnetic action. I was, at one time, of the same opinion, but upon carefully investigating the subject I was led to the conviction that they are more of an electrostatic nature. It must be remembered that in these experiments we have to deal with primary currents of an enormous frequency or rate of change and of high potential, and that the secondary conductor consists of a rarefied gas, and that under such conditions electrostatic effects must play an important part.

In support of my view I will describe a few experiments made by me. To excite luminosity in the tube it is not absolutely necessary that the conductor should be closed. For instance, if an ordinary exhausted tube (preferably of large diameter) be surrounded by a spiral of thick copper wire serving as the primary, a feebly luminous spiral may be induced in the tube, roughly shown in Fig. 3. In one of these experiments a curious phenomenon was observed; namely, two intensely luminous circles, each of them close to a turn of the primary spiral, were formed inside of the tube, and I attributed this phenomenon to the .existence of nodes on the primary. The circles were connected by a faint luminous spiral parallel to the primary and in close proximity to it. To produce this effect I have found it necessary to strain the jar to the utmost. The turns of the spiral tend to close and form circles, but this, of course, would be expected, and does not necessarily indicate an electro-magnetic effect; whereas the fact that a glow can be produced along the primary in the form of an open spiral argues for an electrostatic effect.

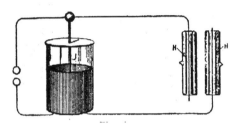

In using Dr. Lodge's recoil circuit, the electrostatic action is likewise apparent. The arrangement is illustrated in Fig. 4. In his experiments two hollow exhausted tubes H H were slipped over the wires of the recoil circuit and upon discharging the jar in the usual manner luminosity was excited in the tubes.

Another experiment performed is illustrated in Fig. 5. In this case an ordinary lamp-bulb was surrounded by one or two turns of thick copper wire P and the luminous circle L excited in the bulb by discharging the jar through the primary. The lamp-bulb was provided with a tinfoil coating on the side opposite to the primary and each time the tinfoil coating was connected to the ground or to a large object the luminosity of the circle was considerably increased. This was evidently due to electrostatic action.

In other experiments I have noted that when the primary touches the glass the luminous circle is easier produced and is more sharply defined; but I have not noted that, generally speaking, the circles induced were very sharply defined, as Prof. J. J. Thomson has observed; on the contrary, in my experiments they were broad and often the whole of the bulb or tube was illuminated; and in one ease I have observed an intensely purplish glow, to which Prof. J. J. Thomson refers. But the circles were always in close proximity to the primary and were considerably easier produced when the latter was very close to the glass, much more so than would be expected assuming the action to be electromagnetic and considering the distance; and these facts speak for an electrostatic effect.

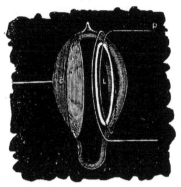

Furthermore I have observed that there is a molecular bombardment in the plane of the luminous circle at right angles to the glass — supposing the circle to be in the plane of the primary — this bombardment being evident from the rapid heating of the glass near the primary. Were the bombardment not at right angles to the glass the hating could not be so rapid. If there is a circumferential movement of the molecules constituting the luminous circle, I have thought that it might be rendered manifest by placing within the tube or bulb, radially to the circle, a thin plate of mica coated with some phosphorescent material and another such plate tangentially to the circle. If the molecules would move circumferentially, the former plate would be rendered more intensely phosphorescent. For want of time I have, however, not been able to perform the experiment.

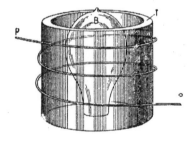

Another observation made by me was that when the specific inductive capacity of the medium between the primary and secondary is increased, the inductive effect is augmented. This is roughly illustrated in Fig. G. In this case luminosity was excited in an exhausted tube or bulb B and a glass tube T slipped between the primary and the bulb, when the effect pointed out was noted. Were the action wholly electromagnetic no change could possibly have been observed.

I have likewise noted that when a bulb is surrounded by a wire closed upon itself and in the plane of the primary, the formation of the luminous circle within the bulb is not prevented. But if instead of the wire a broad strip of tinfoil is glued upon the bulb, the formation of the luminous band was prevented, because then the action was. 'distributed over a greater surface. The effect of the closed tinfoil was no doubt of an electrostatic nature, for it presented a much greater resistance than the closed wire and produced therefore a much smaller electromagnetic effect.

Some of the experiments of Prof. J. J. Thomson also would seem to show some electrostatic action. For instance, in the experiment with the bulb enclosed in a bell jar, I should think that when the latter is exhausted so far that the gas enclosed reaches the maximum conductivity, the formation of the circle in the bulb and jar is prevented because of the space surrounding the primary being highly conducting; when the jar is further exhausted, the conductivity of the space around the primary diminishes and the circles appear necessarily first in the bell jar, as the rarefied gas is nearer to the primary. But were the inductive effect very powerful, they would probably appear in the bulb also. If, however, the bell Jar were exhausted to the highest degree they would very likely show themselves in the bulb only, that is, supposing the vacuous space to be non-conducting. On the assumption that in these phenomena electrostatic actions are concerned we find it easily explicable why the introduction of mercury or the heating of the bulb prevents the formation of the luminous band or shortens the after-glow; and also why in some cases a platinum wire may prevent the excitation of the tube. Nevertheless some of the experiments of Prof. J. J. Thomson would seem to indicate an electromagnetic effect. I may add that in one of my experiments in which a vacuum was produced by the Torricellian method, I was unable to produce the luminous band, but this may have been due to the weak exciting current employed.

My principal argument is the following: I have experimentally proved that if the same discharge which is barely sufficient to excite a luminous band in the bulb when passed through the primary circuit be so directed as to exalt the electrostatic inductive effect — namely, by converting upwards — an exhausted tube, devoid of electrodes, may be excited at a distance of several feet.

Nikola Tesla

Note by Prof. J. J. Thomson

The London Electrician, July 24, 1891

"Mr. Tesla seems to ascribe the effects he observed to electrostatic action, and I have no doubt, from the description he gives of his method of conducting his experiments, that in them electrostatic action plays a very important part. He .seems, however, to have misunderstood my position with respect to the cause of these discharges, which is not, as he implies, that luminosity in tubes without electrodes cannot be produced by electrostatic action, but that it can also be produced when this action is excluded. As a matter of fact, it is very much easier to get the luminosity when these electrostatic effects are operative than when they are not. As an illustration of this I may mention that the first experiment I tried with the discharge of a Leyden jar produced luminosity in the tube, but it was not until after six weeks' continuous experimenting that I was able to get a discharge in the exhausted tube which I was satisfied was due to what is ordinarily called electrodynamic action. It is advisable to have a clear idea of what we mean by electrostatic action. If, previous to the discharge of the jar, the primary coil is raised to a high potential, it will induce over the glass of the tube a distribution of electricity. When the potential of the primary suddenly falls, this electrification will redistribute itself, and may pass through, the rarefied gas and produce luminosity in doing so. Whilst the discharge of the jar is going on, it is difficult, and, from a theoretical point of view, undesirable, to separate the effect into parts, one of which is called electrostatic, the other electromagnetic; what we can prove is that in this case the discharge is not such as would be produced by electromotive forces derived from a potential function. In my experiments the primary coil was connected to earth, and, as a further precaution, the primary was separated from the discharge tube by a screen of blotting paper, moistened with dilute sulphuric acid, and connected to earth. Wet blotting paper is a sufficiently good conductor to screen off a stationary electrostatic effect, though it is not a good enough one to stop waves of alternating electromotive intensity. When showing the experiments to the Physical Society

I could not, of course, keep the tubes covered up, but, unless my memory deceives me, I stated the precautions which had been taken against the electrostatic effect. To correct misapprehension I may state that I did not read a formal paper to the Society, my object being to exhibit a few of the most typical experiments. The account of the experiments in the *Electrician* was from a reporter's note, and was not written, or even read, by me. I have now almost finished writing out, and hope very shortly to publish, an account of these and a large number of allied experiments, including some analogous to those mentioned by Mr. Tesla on the effect of conductors placed near the discharge tube, which I find, in some cases, to produce a diminution, in others an increase, in the brightness of the discharge, as well as some on the effect of the presence of substances of large specific inductive capacity. These seem to me to admit of a satisfactory explanation, for which, however, I must refer to my paper."

N. Tesla

Reply to J. J. Thomson's

The Electrical Engineer - N.Y. — August 26, 1891

In *The Electrical Engineer* of August 12, I find some remarks of Prof. J. J. Thomson, which appeared originally in the London *Electrician* and which have a bearing upon some experiments described by me in your issue of July 1.

I did not, as Prof. J. J. Thomson seems to believe, misunderstand his position in regard to the cause of the phenomena considered, but I thought that in his experiments, as well as in my own, electrostatic effects were of great importance. It did not appear, *from* the meagre description of his experiments, that all possible precautions had been taken to exclude these effects. I did not doubt that luminosity could be excited in a closed tube when electrostatic action is completely excluded. In fact, at the outset, I myself looked for a purely electrodynamic effect and believed that I had obtained it. But many experiments performed at that time proved to me that the electrostatic effects were generally of far greater importance, and admitted of a more satisfactory explanation of most of the phenomena observed.

In using the term *electrostatic* I had reference rather to the nature of the action than to a stationary condition, which is the usual acceptance of the term. To express myself more clearly, I will suppose that near a closed exhausted tube be placed a small sphere charged to a very high potential. The sphere would act inductively upon the tube, and by distributing electricity over the same would undoubtedly produce luminosity (if the potential be sufficiently high), until a permanent condition would be reached. Assuming the tube to be perfectly well insulated, there would be only one instantaneous flash during the act of distribution. This would be due to the electrostatic action simply.

But now, suppose the charged sphere to be moved at short intervals with great speed along the exhausted tube. The tube would now be permanently excited, as the moving sphere would cause a constant redistribution of electricity and collisions of the molecules of the rarefied gas. We would still have to deal with an electrostatic effect, and in addition an electrodynamic effect would be observed. But if it were found that, for instance, the effect produced depended more on the specific inductive capacity than on the magnetic permeability of the medium — which would certainly be the case for speeds incomparably lower than that of light — then I believe I would be justified in saying that the effect produced was more of an electrostatic nature. I do not mean to say, however, that any similar condition prevails in the case of the discharge of a Leyden jar through the primary, but I think that such an action would be desirable.

It is in the spirit of the above example that I used the terms "more of an electrostatic nature," and have investigated the influence of bodies of high specific inductive capacity, and observed, for instance, the importance of the quality of glass of which the tube is made. I also endeavored to ascertain the influence of a medium of high permeability by using oxygen. It appeared from rough estimation that an oxygen tube when excited under similar conditions — that is, as far as could be determined — gives more light; but this, of course, may be due to many causes.

Without doubting in the least that, with the care and precautions taken by Prof. J. J. Thomson, the luminosity excited was due solely to electrodynamic action, I would say that in many experiments I have observed curious instances of the ineffectiveness of the screening, and I have also found that the electrification through the air is often of very great importance, and may, in some cases, determine the excitation of the tube.

In his original communication to the *Electrician,* Prof. J. J. Thomson refers to the fact that the luminosity in a tube near a wire through which a Leyden jar was discharged was noted by Hittorf. I think that the feeble luminous effect referred to has been noted by many experimenters, but in my experiments the effects were much more powerful than those usually noted.

Nikola Tesla

The "Drehstrom" Patent

The Electrical World - N. Y. — Oct. 8, 1892

In the last issue of *the Electrical World* I find an article on my "Drehstrom" patent which appeared originally is *Industries,* and is, I believe, from the pen of the able editor of that journal. Some of the statements made are such as to cause an erroneous opinion to gain ground, which I deem it my duty to prevent — a disagreeable one I may say, as I do not like to express my opinion on a patent, especially if it is my own.

It may be, as the writer states, that the theory of the action of my motor advanced in my paper before the American Institute in May, 1888, is a clumsy one, but this theory was formed by me a number of years before the practical results were announced, the patents being applied for only after it was undoubtedly demonstrated that the motor could fairly compete in efficiency with the direct current motor, and that the invention was one of commercial value. These patents were taken out with the help of some of the ablest attorneys in the United States, well versed in electrical matters; the specifications were drawn up with great care, in view of the importance of the invention, and with proper regard to the state of the art at that period, and had the patents been carefully studied by others there would not hare been various features of my system reinvented, and several inventors, would have been spared at this late date a keen disappointment.

The writer apprehends that it might be difficult for a non-technical judge to decide whether a motor with two or more separate fields and armatures, coupled together mechanically, does or does not fall under my patent. I do not share his apprehension, Judges are highly educated men, and it does not require much technical knowledge to convince one that it is the same whether two belts driving a rigid arbor are close together or far apart. Nor do I think that it is necessary for the honorable judge to be a partisan of the armature reaction theory in order to recognize the identity of the two arrangements referred to by the writer of the article in question. Indeed, I would seriously doubt the sincerity of a man capable of clear conceptions were he to uphold that the arrangements are essentially different, even if the case should stand exactly as he assumes by way of illustration of " puzzles likely to arise." For where is there a difference? Take, for instance, a form of my two-phase motor. There are two sets of field magnets, one at the *neutral* parts of the other. One of the sets, therefore, might as well be removed and placed a distance sideways, but long experience shows that in output, efficiency, cost of construction and in general mechanical respects such an arrangement is inferior. The two sets are connected inductively through the armature body or the windings thereon. Part of the period one set of field magnets acts as a generator, setting up induction currents, which circulate in the field of force of the other set, which may be looked upon as a motor. Part of the period again, the second set becomes the generator and the first the motor, the action being at the same time such that the generated currents are always passed in a definite direction with respect to the field ; they are commutated as it were, and a tendency to rotate in a given direction is imparted to the armature. Now place two fields side by side and connect properly the armature windings. Are not the fields again inductively connected? Do not the currents set up by one field cause currents to circulate in the other, and is the action not exactly the same in both cases? This is a fact, no matter what theory is adhered to. The writer says that in the case of two separate structures there is really nothing which may be called rotation of the field. But is there any such thing, when the two structures are merged in one? Is it not in accordance with accepted notions to conceive the imaginary lines as surging simply in the pole projections in exactly the same manner in both the arrangements? Irrespective of the view taken, be it even the more unfair to the inventor, no one is permitted to go so far as to make him responsible, in such a case, for theories and interpretations of his invention. Theories may come and go, but the motor works, a practical result is achieved and the art is advanced through his pains and efforts. But what I desire to point out principally is that in the article above referred to the writer is only assuming a case which cannot occur. He is evidently judging the state of things from my short paper before the American Institute. This paper was written in a hurry, in fact only shortly before the meeting of the Institute, and I was unable to do full justice even to those features upon which as employee of a company owning the invention, I was permitted to dwell. Allow me to observe that my patent specification was written up more carefully than my paper and the view taken in it is a broader and truer one. While the "clumsy" theory was adapted as the beat in explanation of the action of the motor, the invention is *not* represented as dependent entirely on that theory ; and in showing a three phase motor with six projections, where it was manifestly more consistent with the accepted popular ideas to assume the "lines of force" as simply surging in the projecting pole pieces, this view was distinctly and advisedly taken, as the following quotation from my foundation patent will show : "The variations in the strength and intensity of the currents transmitted through these circuits (lines and armature) and traversing the coils of the motor produce a steadily progressive shifting of the *resultant attractive force* exerted by the poles upon the armature and consequently keep the armature rapidly rotating." There is, in this instance, no question of a rotating field in the common acceptance of the term of the *resultant attractive force* there is a question simply of a diagram of force, and it is immaterial for the operation whether the fields are close together or far apart, or even whether, or not, they are inductively connected.

I do not think that in Germany, where the Patent Office is proverbially strict in upholding the rights of the inventor, an illegitimate and unfair appropriation of the invention by others will be tolerated by the courts.

N. Tesla

The Ewing High-Frequency Alternator and Parsons Steam Engine

Electrician - London — Dec. 17, 1892

In your issue of November 18 I find a description of Prof. Ewing's high-frequency alternator, which has pleased me chiefly because it conveyed to me the knowledge that he, and with him, no doubt, other scientific men, is to investigate the properties of high-frequency currents. With apparatus such as you describe, shortly a number of experimenters, more competent than myself, will be enabled to go over the ground as yet but imperfectly explored, which will undoubtedly result in the observation of novel facts and elimination of eventual errors.

I hope it will not be interpreted as my wishing to detract anything from Prof. Ewing's merit if I state the fact that for a considerable time past I have likewise thought of combining the identical steam turbine with a high-frequency alternator. *Anch'io sono pittore.* I had a number of designs with such turbines, and would have certainly carried them out had the turbines been here easily and cheaply obtainable, and had my attention not been drawn in a different direction. As to the combination to which you give a rather complicated name, I consider it an excellent one. The advantages of using a high speed are especially great in connection with such alternators. When a belt is used to drive, one must resort to extraordinarily large diameters in order to obtain the necessary speed, and this increases the difficulties and cost of construction in an entirely unreasonable proportion. In the machine used in my recent experiments the weight of the active parts is less than 50 pounds, but there is an additional weight of over 100 pounds in the supporting frame, which a very careful constructor would have probably made much heavier. When running at its maximum speed, and with a proper capacity in the armature circuit, two and a one-half horse-power can be performed. The large diameter (30 inches), of course, has the advantage of affording better facility for radiation; but, on the other hand, it is impossible to work with a very small clearance.

I have observed with interest that Prof. Ewing has used a magnet with alternating poles. In my first trials I expected to obtain the best results with a machine of the Mordey type - that is, with one having pole projections of the same polarity. My idea was to energize the field up to the point of the maximum permeability of the iron and vary the induction around that point. But I found that with a very great number of pole projections such a machine would not give good results, although with few projections, and with an armature without iron, as used by Mordey, the results obtained were excellent. Many experiences of similar nature made in the course of my study demonstrate that the ordinary rules for the magnetic circuit do not hold good with high frequency currents. In ponderable matter magnetic permeability, and also specific inductive capacity, must undergo considerable change when the frequency is varied within wide limits. This would render very difficult the exact determination of the energy dissipated in iron cores by very rapid cycles of magnetization, and of that in conductors and condensers, by very quick reversals of current. Much valuable work remains to be done in these fields, in which it is so easy to observe novel phenomena, but so difficult to make quantitative determinations. The results of Prof. Ewing's systematical research will be awaited with great interest.

It is gratifying to note from his tests that the turbines are being rapidly improved. Though I am aware that the majority of engineers do not favor their adoption. I do not hesitate to say that I believe in their success. I think their principle uses, in no distant future, will be in connection with alternate current motors, by means of which it is easy to obtain a constant and, in any desired ratio, reduced speed. There are objections to their employment for driving direct current generators, as the commutators must be a source of some loss and trouble, on account of the very great speed; but with an alternator there is no objectionable feature whatever. No matter how much one may be opposed to the introduction of the turbine, he must have watched with surprise the development of this curious branch of the industry, in which Mr. Parsons-has been a pioneer, and everyone must wish him the success which his skill has deserved.

Nikola Tesla

The Physiological and Other Effects of High Frequency Currents

The Electrical Engineer — Feb. 11, 1893

In *the Electrical Engineer*, of January 25,1893, I note an article by Mr. A. A. C. Swinton, referring to my experiments with high frequency currents. Mr: Swinton uses in these experiments the method of converting described by me in my paper before the American Institute of Electrical Engineers, in May, 1891, and published in *the Electrical Engineer* of July 8, 1891, which has since been employed by a number of experimenters; but it has somewhat surprised me to observe that he makes use of an ordinary vibrating contact-breaker, whereas he could have employed the much simpler method of converting continuous currents into alternating currents of any frequency which was shown by me two years ago. This method does not involve the employment of any moving parts, and allows the experimenter to vary the frequency et will by simple adjustments. I had thought that most electricians were at present familiar with this mode of conversion which possesses many beautiful features.

The effects observed by Mr. Swinton are not new to rue and they might have been anticipated by those who have carefully read what I !rave written on the subject. But I cannot agree with some of the views expressed by him.

First of all, in regard to the physiological effects. I have made a clear statement at the beginning of my published studies, and my continued experience :with these currents has only further strengthened me is the opinion then expressed. I stated in my paper, before mentioned, that it is an undeniable fact that currents of very high frequency are less injurious than the low frequency currents, :but I have also taken care to prevent the idea from gaining ground that these currents are absolutely harmless, as will be evident, from the following quotation: " If received directly from a machine or from a secondary of low resistance, they (high frequency currents) produce more or less powerful effects, and may, cause serious injury, especially when used in conjunction with condensed." This refers to currents of ordinary' potential differences such as are used in general commercial practice.

As regards the currents of very high potential differences, which were employed an my, experiments, I have never considered the current's strength, but the energy which the human body eras capable of receiving without injury, and I have expressed this quite clearly on more than one occasion. For instance, I stated teat "the higher, the frequency the greater the amount of electrical energy which may be passed through the body without serious discomfort." And on another occasion when a high tension coil was short-circuited though the body of the experimenter I stated that the immunity was due to the fact that less energy was available externally to the coil when the experimenter's body joined the terminals. This is practically what Mr. Swinton expresses in another way; namely, by saying that with "high frequency currents it is possible to obtain effects with exceedingly small currents," etc.

In regard to the experiments, with lamp filaments, I have, I believe, expressed myself with equal clearness. I have pointed out come phenomena of impedance which at that time (1891) were considered very striating, and I have also pointed out the great importance of the rarefied gas surrounding the filament when we have to deal with currents of such high frequency. The heating of the filament by s comparatively small current is not, as Mr. Swinton thinks, due to its impedance or increased ohmic resistance, but principally to the presence of rarefied gas in the bulb. Ample evidence of the truth of this can be obtained in vary many experiments, and to cite them would be merely lengthening this communication unduly.

Likewise, observations made when the experimenter's body was included in the path of the discharge, are, in my opinion, not impedance, but capacity, phenomena. The spark between the hands is the shorter, the larger the surface of the body, and no spark whatever would, be, obtained if the surface of the body were sufficiently large.

I would here point out that one is apt to fall into the error of supposing that the spark which is produced between two points on a conductor, not very distant from each other, is due to the impedance of the conductor. This is certainly the case when the current is of considerable strength. But when there is a vibration along a wire which is constantly maintained, and the current is inappreciable whereas the potential at the coil terminal is exceedingly high, then lateral dissipation comes into play prominently. There is then owing to this dissipation, a rapid fall of potential along the wire and high potential differences may exist between points only a short distance apart This is of course not to be confounded with those differences of potential observed between points when there are fixed wares with ventral and nodal points maintained on a conductor. The lateral dissipation, and not the skin effect, is, I think, the reason why so great an amount of energy may be passed into the body of a person without causing discomfort.

It always affords me great pleasure to note; that something which I have suggested is being employed for some instructive or practical purpose; but I may be pardoned for mentioning that other observations made by Mr. Swinton, and by other experimenters, have recently been brought forward as novel, and arrangements of apparatus which I have suggested have been used repeatedly by some who apparently are in complete ignorance of what I have done in this direction.

Nikola Tesla

From Nikola Tesla - He Writes about His Experiments in Electrical Healing

The Detroit Free Press — Feb. 16, 1896

Some weeks ago this journal published an interesting article concerning electrical oscillations as observed by the eminent scientist, Nicola Tesla. So much interest was shown in the subject that Mr. Tesla was appealed to directly and in response to that appeal he sends to *The Detroit Free Press* this open letter:

Nos. 46 & 48 E. Houston Street
New York, February 10, 1896

During the past few weeks I have received so many letters concerning the same subject that it was entirely beyond my power to answer all of them individually. In view of this I hope that I shall be excused for the delay, which I must regret, in acknowledging the receipt, and also for addressing this general communication in answer to all inquiries.

The many pressing demands which have been made upon me in consequence of exaggerated statements of the journals have painfully impressed me with the fact that there are a great many sufferers, and furthermore that nothing finds a more powerful echo than a promise held out to improve the condition of the unfortunate ones.

The members of the medical fraternity are naturally more deeply interested in the task of relieving the suffering from their pain, and, as might be expected, a great many communications have been addressed to me by physicians. To these chiefly this brief statement of the actual facts is addressed.

Some journals have confounded the physiological effects of electrical oscillations with those of mechanical vibrations, this being probably due to the circumstance that a few years ago I brought to the attention of the scientific men some novel methods and apparatus for the production of electrical oscillations which, I learn, are now largely used in some modification or other in electro-therapeutic treatment and otherwise. To dispel this erroneous idea I wish to state that the effects of purely mechanical vibrations which I have more recently observed, have nothing to do with the former.

Mechanical vibrations have often been employed locally with pronounced results in the treatment of diseases, but it seems that the effects I refer to have either not been noted at all, or if so, only to a small degree, evidently because of the insufficiency of the means which have eventually been employed in the investigations.

While experimenting with a novel contrivance, constituting in its simplest form a vibrating mechanical system, in which from the nature of the construction the applied force is always in resonance with the natural period, I frequently exposed my body to continued mechanical vibrations. As the elastic force can be made as large as desired, and the applied force used be very small, great weights, half a dozen persons, for instance, may be vibrated with great rapidity by a comparatively small apparatus.

I observed that such intense mechanical vibrations produce remarkable physiological effects. They affect powerfully the condition of the stomach, undoubtedly promoting the process of digestion and relieving the feeling of distress, often experienced in consequence of the imperfect function of the organs concerned in the process. They have a strong influence upon the liver, causing it to discharge freely, similarly to an application of a catharic. They also seem to affect the glandular system, notable in the limbs; also the kidneys and bladder, and more or less influence the whole body. When applied for a longer period they produce a feeling of immense fatigue, so that a profound sleep is induced.

The excessive tiring of the body is generally accompanied by nervous relaxation, but-there seems to be besides a specific action on the nerves.

These observations, though incomplete, are, in my own limited judgment, nevertheless positive and unmistakable, and in view of this and of the importance of further investigation of the subject by competent men I prepared about a year ago a machine with suitable adjustments for varying the frequency and amplitude of the vibrations, intending to give it to some medical faculty for investigation. This machine, together with other apparatus, was unfortunately destroyed by fire a year ago, but will be reconstructed as soon as possible.

In making the above statements I wish to disconnect myself with the extraordinary opinions expressed in some journals which I have never authorized and which, though they may have been made with good intent, cannot fail to be hurtful by giving rise to visionary expectations.

Yours very truly,

N. Tesla

Tesla's Latest Results: He Now Produces Radiographs At A Distance of More Than Forty Feet

Electrical Review - N. Y. — *March 18, 1896*

To The Editor of Electrical Review:

Permit me to say that I was slightly disappointed to note in your issue of Mar. 11 the prominence you have deemed to accord to my youth and talent, while the ribs and other particulars of which, with reference to the print accompanying my communication, I described as clearly visible, were kept modestly in the background. I also regretted to observe an error in one of the captions, the more so, as I must ascribe it to my own text. I namely stated on page 135, third column, seventh line: "A similar impression was obtained through the body of the experimenter, etc., through a distance of four feet." The general truth of the fact of taking such a shadow at the distance given is concerned, your caption might as well stand, for I am producing strong shadows at distances of 40 feet. I repeat, 40 feet and even more. Nor is this all. So strong are the actions on the film that provisions must be made to guard the plates in my photographic department, located on the floor above, a distance of fully 60 feet, from being spoiled by long exposure to the stray rays. Though during my investigations I have performed many experiments which seemed extraordinary, I am deeply astonished observing these unexpected manifestations, and still more so, as even now I see before me the possibility, not to say certitude, of augmenting the effects with my apparatus at least tenfold! What may we then expect? We have to deal here, evidently, with a radiation of astonishing power, and the inquiry into its nature becomes more and more interesting and important. Here is an unlooked-for result of an action which, though wonderful in itself, seemed feeble and entirely incapable of such expansion, and affords a good example of the fruitfulness of original discovery. These effects upon the sensitive plate at so great a distance I attribute to the employment of a bulb with a single terminal, which permits the use of practically any desired potential and the attainment of extraordinary speeds of the projected particles. With such a bulb it is also evident that the action upon a fluorescent screen is proportionately greater than when the usual kind of tube is employed, and I have already observed enough to feel sure that great developments are to be looked for in this direction. I consider Roentgen's discovery, of enabling us to see, by the use of a fluorescent screen, through an opaque substance, even a more beautiful one than the recording upon the plate.

Since my previous communication to you I have made considerable progress, and can presently announce one more result of importance. I have lately obtained shadows by reflected rays only, thus demonstrating beyond doubt that the Roentgen rays possess this property. One of the experiments may be cited here. A thick copper tube, about a foot long, was taken and one of its ends tightly closed by the plateholder containing a sensitive plate, protected by a fiber cover as usual. Near the open end of the copper tube was placed a thick plate of glass at an angle of 45 degrees to the axis of the tube. A single-terminal bulb was then suspended above the glass plate at a distance of about eight inches, so that the bundle of rays fell upon the latter at an angle of 45 degrees, and the supposedly reflected rays passed along the axis of the copper tube. An exposure of 45 minutes gave a clear and sharp shadow of a metallic object. This shadow was produced by the reflected rays, as the direct action was absolutely excluded, it having been demonstrated that even under the severest tests with much stronger actions no impression whatever could be produced upon the film through a thickness of copper equal to that of the tube. Concluding from the intensity of the action by comparison with an equivalent effect due to the direct rays, I find that approximately two per cent of the latter were reflected from the glass plate in this experiment. I hope to be able to report shortly and more fully on this and other subjects.

In my attempts to contribute my humble share to the knowledge of the Roentgen phenomena, I am finding more and more evidence in support of the theory of moving material particles. It is not my intention, however, to advance at present any view as to the bearing of such a fact upon the present theory of light, but I merely seek to establish the fact of the existence of such material streams in so far as these isolated effects are concerned. I have already a great many indications of a bombardment occurring outside of the bulb, and I am arranging some crucial tests which, I hope, will be successful. The calculated velocities fully account for actions at distances of as much as 100 feet from the bulb, and that the projection through the glass takes place seems evident from the process of exhaustion, which I have described in my previous communication. An experiment which is illustrative in this respect, and which I intended to mention, is the following; If we attach a fairly exhausted bulb containing an electrode to the terminal of a disruptive coil, we observe small streamers breaking through the side of the glass. Usually such a streamer will break through the seal and crack the bulb, whereupon the vacuum is impaired; but, if the seal is placed above the terminal, or if some other provision is made to prevent the streamer from passing through the glass at that point, it often occurs that the stream breaks out through the side of the bulb, producing a fine hole. Now, the extraordinary thing is that, in spite of the connection to the outer atmosphere, the air can not rush into the bulb as long as the hole 'is very small. The glass at the place where the rupture has occurred may grow very hot — to such a degree as to soften; but it will not collapse, but rather bulge out, showing that a pressure from the inside greater than that of the atmosphere exists. On frequent occasions I have observed that the glass bulges out and the hole, through which the streamer rushes out, becomes so large as to be perfectly discernible to the eye. As the matter is expelled from the bulb the rarefaction increases and the streamer becomes less and less intense, whereupon the glass closes again, hermetically sealing the opening. The process of rarefaction, nevertheless, continues, streamers being still visible on the heated place until the highest degree of exhaustion is reached, whereupon they may disappear. Here, then, we have a positive evidence that matter is being expelled through the walls of the glass.

When working with highly strained bulbs I frequently experience a sudden, and sometimes even painful, shock in the eye. Such shocks may occur so often that the eye gets inflamed, and one can not be considered over-cautious if he abstains from watching the bulb too closely. I see in these shocks a further evidence of larger particles being thrown off from the bulb.

Nikola Tesla.

New York, March 14.

Mr. Tesla on Thermo Electricity

The Electrical Engineer - N. Y. — *December 23, 1896*

In a letter to the editor of the Buffalo Enquirer, Mr. Nikola Tesla replies as follows in regard to an inquiry on the subject of the future of electricity:

"The transmission of power has interested me not only as a technical problem, but far more in its bearing upon the welfare of mankind. In this sense I have expressed myself in a lecture, delivered some time ago.

"Since electrical transmission of energy is a process much more economical than any other we know of, it necessarily must play an important part in the future, no matter how the primary energy is derived from the sun. Of all the ways the utilization of a waterfall seems to be the simplest and least wasteful. Even if we could, by combining carbon in a battery, convert the work of the chemical combination into electrical energy with very high economy, such mode of obtaining power would, in my opinion, be no more than a mere makeshift, bound to be replaced sooner or later by a more perfect method, which implies no consumption of any material whatever."

N. Tesla

Tesla's Latest Advances in Vacuum-tube Lighting. Application of Tubes of High Illuminating Power to Photography and Other Purposes

Electrical Review - N. Y. — *Jan. 5, 1898*

To the Editor of *Electrical Review*:

A few years ago I began a series of experiments with a view of ascertaining the applicability of the light emitted by phosphorescent vacuum tubes to ordinary photography. The results soon showed that, even with a tube giving no more light than the equivalent of one half of a candle, objects could be easily photographed with exposures of a few minutes, and the time could be reduced at will by pushing the tube to a high candlepower. Photographs of persons were likewise obtained at that time and, if I am not mistaken, these were the first likenesses produced with this kind of illumination. However, a number of facts, not pertaining to the subject presently considered, were observed in the course of the experiments which, had they been immediately published, might have materially hastened important scientific developments which have taken place since. To dwell on these and other experimental results obtained at that time, more extensively at the first opportunity, is one of my good resolutions for the coming year. A calamity unfortunately, interrupted my labors for a short period, but as soon as I was able I took up again the thread of the investigation, which was not only interesting in connection with the principal object in view, but was also useful in many other respects. So, for instance, in making observations as to the efficiency or any peculiarity of the vacuum tubes, the photographic plate was found to be an excellent means of comparison, note being taken of the distance and time of exposure, character of the phosphorescent body, degree of rarefaction and other such particulars of the moment.

A rather curious feature in the photographs obtained with tubes of moderate illuminating power, as a few candles, was that the lights and shadows came out remarkably strong, as when very short exposures are made by flashlight, but the outlines were not sharp and practically no detail was visible. By producing tubes of much greater candlepower, a notable improvement in this respect was effected, and this advance prompted me to further efforts in this direction, which finally resulted in the production of a tube of an illuminating power of equal to that of hundreds, and even thousands, of ordinary vacuum tubes. What is more, I believe that I am far from having attained the limit in the amount of light producible, and believe that this method of illumination will be eventually employed for lighthouse purposes. This probably will be considered the oddest and most unlooked-for development of the vacuum tube.

Simultaneously with this progress a corresponding improvement was made in the efficiency of the light produced. A few words on this point might not be amiss, considering that a popular and erroneous opinion still exists in regard to the power consumed by vacuum tubes lighted by ordinary means. So deeply rooted is this opinion which, I will frankly confess, I myself shared for a long time, that, shortly after my own first efforts, Sir David Solomons and Messrs. Pike & Harris undertook to introduce in England such tubes on a large scale in competition with the incandescent system of lighting. The enterprise, which was commented on in the technical periodicals, was commendable enough, but it was not difficult to foretell its fate; for although the high-frequency currents obtained from the alternator yielded better economical results than interrupted currents, and although they were obtained in a convenient and fairly economical manner, still the efficiency of the whole system was necessarily too small for competition with incandescent lamps. The reason for the great power consumption, which may often be as much as 10 times that taking place in incandescent lamps for an equivalent amount of light, are not far to seek. A vacuum tube, particularly if it be very large, offers an immense radiating surface, and is capable of giving off a great amount of energy without rising perceptibly in temperature. What still increases the dissipation of energy is the high temperature of the rarefied gas. Generally it is supposed that the particles are not brought to a high temperature, but a calculation from the amount of matter contained in the tube, leads to results which would seem to indicate that, of all the means at disposal for bringing a small amount of matter to a high temperature, the vacuum tube is the most effective. This observation may lead to valuable uses of such tubes in astronomical researches, and a line of experiment to this end was suggested to me recently by Dr. Geo. E. Hale, of the Yerkes Observatory. As compared with these disadvantages the incandescent lamp, crude and inefficient as it undoubtedly is, possesses vastly superior features. These difficulties have been recognized by me early, and my efforts during the past few years have been directed towards overcoming these defects and have finally resulted in material advances, so that I find it possible to obtain from a tube of a volume not much greater than that of a bulb of an incandescent lamp, about the same amount of light produced by the latter, without the tube becoming overheated, which is sure to take place under ordinary conditions. Both of these improvements, the increase of candle-power as well as degree of efficiency, have been achieved by gradual perfection of the means of producing economically harmonical electrical vibrations of extreme rapidity. The fundamental principle involved is now well known, and it only remains to describe the features of the system in detail, a duty with which I expect to be able to comply soon, this being another one of my good resolutions.

The purpose of the present communication is chiefly to give an idea in how far the object here aimed at was obtained. The photographs shown were taken by a tube having a radiating surface of about two hundred square inches. The frequency of the oscillations, which were obtained from an Edison direct-current supply circuit, I estimated to be about two million a second. The illuminating power of the tube approximated about one thousand candles, and the exposures ranged from two to five seconds, the distance of the object being four to five feet from the tube. It might be asked why, with so high an illuminating power, the exposures should not be instantaneous. I would not undertake to satisfactorily answer this question, which was put to me recently by a scientific man, whose visit to my laboratory I still vividly recollect. Likenesses can, of course, be obtained with instantaneous exposures, but it has been found preferable to expose longer and at a greater distance from the tube. The results so far obtained would make it appear that this kind of light will be of great value in photography, not only because the artist will be able to exactly adjust the conditions in every experiment so as to secure the best result, which is impossible with ordinary light. He will thus be made entirely independent of daylight, and will be able to carry on his work at any hour, night or day. It might also be of value to the painter, though its use for such purposes I still consider problematical.

Nikola Tesla

New York, Jan. 3

Tesla on Animal Training by Electricity
New York Journal — Feb. 6, 1898

To the Editor of the *Journal*:

It seems to me that there are interesting possibilities in the training of animals by electricity. Of course,. it's rather out of my province, but the idea of the electrical subjugator appears feasible when one knows the power of electricity and the instinctive fear that brutes have of the unknown. And the electrical method seems more humane than those I believe are in use - the whip, red hot irons, and drugs, which are likely to do permanent injury, while the physical effects of an electric shock are soon gone, only the moral ones remaining.

The subjugator referred to will do the work, but I think an apparatus could be designed that would be less dangerous to the man. I do not desire to be understood as giving the matter deep thought, but believe that if instead of the armored backpad, the trainer used a wand, with two prongs at one end, better results would follow. This wand would be connected with the supply cables and could be applied to any part of the animal's body at will. Its operation would be precisely the same as the subjugator here illustrated, the two prongs supplying the positive and negative poles of contact found in the flattened wires. With this wand an animal could be simply shocked, stunned or killed, as required.

To cure animals of jumping at men in cages, a screen of stout but flexible wire could be stretched between the trainer and his subject, the wires to be alternately positive and negative, and connected through the regulator with the dynamo. After a couple of springs which would hurl him half insensible back into his corner, the taste for unexpected jumps would leave the brute.

[The following article appeared with the above.-Ed]
Prague, Jan. 22.

Science has come to aid the lion tamer in subduing the wild beast. The red hot iron will, in future, be cast aside as unnecessary and out of date. Live wires, surcharged with electricity that baffle the lion's fiercest assaults, and burn and maim him badly have taken the place of the lash and scorching iron. A lion tamer of

Austria, Louis Koemmenich, has been the first to call in the assistance of the lightning to subdue wild beasts.

Koemmenich has invented what he calls the electrical subjugator. This is a shield of electric wires that fasten on the back of the lion tamer and are connected with a dynamo by a wire coil of sufficient length to allow Koemmenich to move around the cage.

In his hand he will carry a charged metal ball on an insulated handle, to be used as the red hot iron was in former days.

The dynamo is operated by an assistant outside of the cage.

Should a lion show a disposition to leap on Koemmenich, he invites attack by deliberately turning his back to the lion and apparently encouraging the onslaught. When the beast springs his paws come in contact with the electric shield, and he receives a shock of 1,500 volts from the dynamo.

The operator can, if necessary, increase the voltage so as to shock the animal to death.

Thus far the device has worked like magic. One dose of lightning is sufficient for the average lion. Whips and even hot irons they have dared, but no animal has yet troubled Koemmenich after receiving into its body 1,500 volts from the electric subjugator. Whenever Koemmenich enters the cage after an encounter with a lion that has run against the electrical subjugator, he will cower away into a corner of the cage, and never need any further punishment.

Nikola Tesla

Letter to Editor

Electrical Engineer - N. Y. — *Nov. 24, 1898*

New York, Nov. 18, 1898
46 & 48 East Houston St.

Editor of The Electrical Engineer, 120 Liberty St., New York City:
Sir - By publishing in your columns of Nov. 17 my recent contribution to the Electro-Therapeutic Society you have finally succeeded - after many vain attempts made during a number of years - in causing me a serious injury. It has cost me great pains to write that paper, and I have expected to see it appear among other dignified contributions of its kind, and I confess, the wound is deep. But you will have no opportunity for inflicting a similar one, as I propose to take better care of my papers in the future. In what manner you have secured this one in advance of other electrical periodicals who had an equal right to the same, rests with the secretary of the society to explain.

Your editorial comment would not concern me in the least, were it not my duty to take note of it. On more than one occasion you have offended me, but in my qualities both as Christian and philosopher I have always forgiven you and only pitied you for your errors. This time, though, your offence is graver than the previous ones, for you have dared to cast a shadow on my honor.

No doubt you must have in your possession, from the illustrious men whom you quote, tangible proofs in support of your statement reflecting on my honesty. Being a bearer of great honors from a number of American universities, it is my duty, in view of the slur thus cast upon them, to exact from you that in your next issue you produce these, together with this letter, which in justice to myself, I am forwarding to other electrical journals. In the absence of such proofs, I require that, together with the preceding, you publish instead a complete and humble apology for your insulting remark which reflects on me as well as on those who honor me.

On this condition I will again forgive you; but I would advise you to limit yourself in your future attacks to statements for which you are not liable to be punished by law.
N. Tesla

Tesla Describes His Efforts in Various Fields of Work

The Sun, New York, November 21, 1898

To the Editor of the Sun

Sir: Had it not been for other urgent duties, I would before this have acknowledged your highly appreciative editorial of November 13. Such earnest comments and the frequent evidences of the highest appreciation of my labors by men who are the recognized leaders of this day in scientific speculation, discovery and invention are a powerful stimulus, and I am thankful for them. There is nothing that gives me so much strength and courage as the feeling that those who are competent to judge have faith in me.

Permit me on this occasion to make a few statements which will define my position in the various fields of investigation you have touched upon.

I can not but gratefully acknowledge my indebtedness to earlier workers, as Dr. Hertz and Dr. Lodge, in my efforts to produce a practical and economical lighting system on the lines which I first disclosed in a lecture at Columbia College in 1891. There exists a popular error in regard to this light, inasmuch as it is believed that it can be obtained without generation of heat. The enthusiasm of Dr. Lodge is probably responsible for this error, which I have pointed out early by showing the impossibility of reaching a high vibration without going through the lower or fundamental tones. On purely theoretical grounds such a result is thinkable, but it would imply a device for starting the vibrations of unattainable qualities, inasmuch as it would have to be entirely devoid of inertia and other properties of matter. Though I have conceptions in this regard, I dismiss for the present this proposition as being impossible. We can not produce light without heat, but we can surely produce a more efficient light than that obtained in the incandescent lamp, which, though a beautiful invention, is sadly lacking in the feature of efficiency. As the first step toward this realization, I have found it necessary to invent some method for transforming economically the ordinary currents as furnished from the lighting circuits into electrical vibrations of great rapidity. This was a difficult problem, and it was only recently that I was able to announce its practical and thoroughly satisfactory solution. But this was not the only requirement in a system of this kind. It was necessary also to increase the intensity of the light, which at first was very feeble. In this direction, too, I met with complete success, so that at present I am producing a thoroughly serviceable and economical light of any desired intensity. I do not mean to say that this system will revolutionize those in use at present, which have resulted from the cooperation of many able men. I am only sure that it will have its fields of usefulness.

As to the idea of rendering the energy of the sun available for industrial purposes, it fascinated me early but I must admit it was only long after I discovered the rotating magnetic field that it took a firm hold upon my mind. In assailing the problem I found two possible ways of solving it. Either power was to be developed on the spot by converting the energy of the sun's radiations or the energy of vast reservoirs was to be transmitted economically to any distance. Though there were other possible sources of economical power, only the two solutions mentioned offer the ideal feature of power being obtained without any consumption of material. After long thought I finally arrived at two solutions, but on the first of these, namely, that referring to the development of power in any locality from the sun's radiations, I can not dwell at present. The system of power transmission without wires, in the form in which I have described it recently, originated in this manner. Starting from two facts that the earth was a conductor insulated in space, and that a body can not be charged without causing an equivalent displacement of electricity in the earth, I undertook to construct a machine suited for creating as large a displacement as possible of the earth's electricity.

This machine was simply to charge and discharge in rapid succession a body insulated in space, thus altering periodically the amount of electricity in the earth, and consequently the pressure all over its surface. It was nothing but what in mechanics is a pump, forcing water from a large reservoir into a small one and back again. Primarily I .contemplated only the sending of messages to great distances in this manner, and I described the scheme in detail, pointing out on that occasion the importance of ascertaining certain electrical conditions of the earth. The attractive feature of this plan was that the intensity of the signals should diminish very little with the distance, and, in fact, should not diminish at all, if it were not for certain losses occurring, chiefly in the atmosphere. As all my previous ideas, this one, too, received the treatment of Marsyas, but it forms, nevertheless, the basis of what is now known as "wireless telegraphy." This statement will bear rigorous examination, but it is not made with the intent of detracting from the merit of others. On the contrary, it is with great pleasure that I acknowledge the early work of Dr. Lodge, the brilliant experiments of Marconi, and of a later experimenter in this line, Dr. Slaby, of Berlin. Now, this idea I extended to a system of power transmission, and I submitted it to

Helmholtz on the occasion of his visit to this country. He unhesitatingly said that power could certainly be transmitted in this manner, but he doubted that I could ever produce an apparatus capable of creating the high pressures of a number of million volts, which were required to attack the problem with any chance of success, and that I could overcome the difficulties of insulation. Impossible as this problem seemed at first, I was fortunate to master it in a comparatively short time, and it was in perfecting this apparatus that I came to a turning point in the development of this idea. I, namely, at once observed that the air, which is a perfect insulator for currents produced by ordinary apparatus, was easily traversed by currents furnished by my improved machine, giving a tension of something like 2,500,000 volts. A further investigation in this direction led to another valuable fact; namely, that the conductivity of the air for these currents increased very rapidly with its degree of rarefaction, and at once the transmission of energy through the upper strata of air, which, without such results as I have obtained, would be nothing more than a dream, became easily realizable. This appears all the more certain, as I found it quite practicable to transmit, under conditions such as exist in heights well explored, electrical energy in large amounts. I have thus overcome all the chief obstacles which originally stood in the way, and the success of my system now rests merely on engineering skill.

Referring to my latest invention, I wish to bring out a point which has been overlooked. I arrived, as has been stated, at the idea through entirely abstract speculations on the human organism, which I conceived to be a self-propelling machine, the motions of which are governed by impressions received through the eye. Endeavoring to construct a mechanical model resembling in its essential, material features the human body, I was led to combine a controlling device, or organ sensitive to certain waves, with a body provided with propelling and directing mechanism, and the rest naturally followed. Originally the idea interested me only from the scientific point of view, but soon I saw that I had made a departure which sooner or later must produce a profound change in things and conditions presently existing. I hope this change will be for the good only, for, if it were otherwise, I wish that I had never made the invention. The future may or may not bear out my present convictions, but I can not refrain from saying that it is difficult for me to see at present how, with such a principle brought to great perfection, as it undoubtedly will be in the course of time, guns can maintain themselves as weapons. We shall be able, by availing ourselves of this advance, to send a projectile at much greater distance, it will not be limited in any way by weight or amount of explosive charge, we shall be able to submerge it at command, to arrest it in its flight, and call it back, and to send it out again and explode it at will, and, more than this, it will never make a miss, since all chance in this regard, if hitting the object of attack were at all required, is eliminated. But the chief feature of such a weapon is still to be told; namely, it may be made to respond only to a certain note or tune, it may be endowed with selective power. Directly such an arm is produced, it becomes almost impossible to meet it with a corresponding development. It is this feature, perhaps, more than in its power of destruction, that its tendency to arrest the development of arms and to stop warfare will reside. With renewed thanks, I remain,

Very truly, yours,

N. Tesla

New York,

November 19

Electrical oscillator activity ten million Horsepower Burning atmospheric nitrogen by high frequency discharges twelve million volts.
January 1. 1904

I wish to announce that in connection with the commercial introduction of my inventions I shall render professional services in the general capacity of consulting electrician and engineer.

The near future, I expect with confidence, will be a witness of revolutionary departures in the production, transformation and transmission of energy, transportation, lighting, manufacture of chemical compounds, telegraphy, telephony and other arts and industries.

In my opinion, these advances are certain to follow from the universal adoption of high-potential and high-frequency currents and novel regenerative processes of refrigeration to very low temperatures.

Much of the old apparatus will have to be improved, and much of the new developed, and I believe that while furthering my own inventions, I shall be more helpful in this evolution by placing at the disposal of others the knowledge and experience I have gained.

Special attention will be given by me to the solution of problems requiring both expert information and inventive resource — work coming within the sphere of my constant training and predilection.

I shall undertake the experimental investigation and perfection of ideas, methods and appliances, the devising of useful expedients and, in particular, the design and construction of machinery for the attainment of desired results.

Any task submitted to and accepted by me, will be carried out thoroughly and conscientiously.

Nikola Tesla
Laboratory, Long Island, N. Y.
Residence, Waldorf, New York City

Letter from Nikola Tesla

New York Sun — November 27, 1904

My attention has been called to numerous comments on my letter, published in your issue of November 1, and relating to the electrical equipment of the newly opened catacomb in this city. Some of them are based on erroneous assumptions, which it is necessary for me to correct.

When I stated that my system was adopted, I did not mean that I originated every electrical appliance in the subway. For instance, the one which that ill fated electrician was repairing when he was killed, two days after the catacomb was ready for public use, was not invented by me. Nor was that other device on the sidetracked car, which, as will be remembered, caused the burning of two men. I also must deny any connection with that switch or contrivance which was responsible for the premature death of a man immediately afterward, as well as with that other, which cut short the life of his unfortunate successor. None of these funeral devices, I emphatically state, or any of the other which brought on collisions, delays and various troubles and were instrumental in the loss of arms and legs of several victims, are of my invention, nor do they form, in my opinion, necessary appurtenances of an intelligently planned scheme for the propulsion of cars. Referring to these contrivances, it is significant to read in some journals of the 8th inst. that a small firm failed because their bid was too low. This is indicative of keen competition and sharp cutting of prices, and does not seem in keeping with the munificence claimed for the Interborough Company.

I merely intended to say in my letter that my system of power transmission with three-phase generators and synchronous motor converters was adopted in the subway, the same as on the elevated road. I devised it many years ago for the express purpose of meeting the varied wants of a general electrical distribution of light and power. It has been extensively introduced all over the world because of its great flexibility, and under such conditions of use has been found of great value. But the idea of employing in this great city's main artery, in a case presenting such rigid requirements, this flexible system, offering innumerable chances for breakdowns, accidents and injuries to life and property, is altogether too absurd to dignify it with any serious comment. Here only my multiphase system, with induction motors and closed coil armatures - apparatus unfailing in its operation and minimizing the dangers of travel - should have been installed. Nothing, not even ignorance, will prevent its ultimate adoption; and the sooner the change is made the better it will be for all concerned. Personally, I have no financial or other interest in the matter, except that as a long resident of this city I would have been glad to see my inventions properly used to the advantage of the community. Under the circumstances I must forego this gratification.

The consequences of the unpardonable mistake of the Interborough Company are not confined to this first subway or even to this city. We are driven to travel underground. The elevated road is the eighth wonder, as colossal and imposing in the feature of public forbearance as the Pyramid of Cheops in its dimensions. Sooner or later all interurban railways must be transformed into subterranean. This will call for immense investments of capital, and if defective electrical apparatus is generally adopted the damage to life and property will be incalculable, not to speak of inconvenience to the public.

It seems proper to me to acknowledge on this occasion the painstaking suggestions of some friends of mine, mostly unknown to me, both in the large domain of electrical achievement and in the small sphere of my friendship, to again address the American Institute. It is customary with scientific men to present an original subject only once. I have done so and do not desire to depart from this established precedent. A lecture on the defects of the subway offers great opportunities, but would not be original. In view of certain insinuations I may cite a recently published statement of Mr. C. F. Scott, formerly president of the American Institute: 'n a matter of history it is the Tesla principle and the Tesla system which have been the directing factors in modern electrical engineering practice.' There are but a few men whose acknowledgment of my own work I would quote. Mr. Scott is one of them, as the man whose co-operation was most efficient in bringing about the great industrial revolution through these inventions. But the suggestions of my good friends have fallen on fruitful ground, and should it be possible for me to spare time and energy I may ask the city authorities for power to investigate the subway, and make a sworn report to them on all the defects and deficiencies I may discover, in the interest of public welfare.

A few more words in relation to the signs. With all due respect to general opinion, I entertain quite a different view on that subject. Advertising is a useful art, which is being lifted continually to a higher plane, and will soon be quite respectable. It should not be hampered, but rather encouraged. I would give the Interborough Company every facility for exploiting it, restricting it only in so far as the artistic execution is concerned. A commission of capable men comprising a painter, a sculptor, an architect, a literary man, an engineer and an executive business man might be appointed, to pass upon the merits of the signs submitted for acceptance. I do not see why the public should object to them if they were regulated in this manner. They will further business, make travel less tedious, and help many skillful artisans. The subways are bound to become municipal property, and the city will then derive a revenue from them. What is most important for the safety of life and property, quickness and security of travel, should be first considered. All this depends on the electrical equipment. The engineers have built a good tunnel, and proper apparatus should be installed to match it.

Nikola Tesla
New York, Nov. 26

Electric Autos
Manufacturers' Record — Dec. 29, 1904

Mr. Albert Phenis, Special Correspondent Manufacturers' Record, New York:

Dear Sir - Replying to your inquiry of yesterday, the application of electricity to the propulsion of automobiles is certainly a rational idea. I am glad to know that Mr. Lieb has undertaken to put it into practice. His long experience with the General Electric Co. and other concerns must have excellently fitted him for the task.

There is no doubt that a highly-successful machine can be produced on these lines. The field is inexhaustible, and this new type of automobile, introducing electricity between the prime mover and the wheels, has, in my opinion, a great future.

I have myself for many years advocated this principle. Your will find in numerous technical publications statements made by me to this effect. In my article in the Century, June, 1900, I said, in dealing with the subject: 'Steamers and trains are still being propelled by the direct application of steam power to shafts or axles. A much greater percentage of the heat energy of the fuel could be transformed in motive energy by using, in place of the adopted marine engines and locomotives, dynamos driven by specially designed high-pressure steam or gas engines, by utilizing the electricity generated for the propulsion. Again of 50 to 100 percent, in the effective energy derived from the fuel could be secured in this manner. It is difficult to understand why a fact so plain and obvious is not receiving more attention from engineers.

At first glance it may appear that to generate electricity by an engine and then apply the current to turn a wheel, instead of turning it by means of some mechanical connection with the engine, is a complicated and more or less wasteful process. But it is not so; on the contrary, the use of electricity in this manner secures great practical advantages. It is but a question of time when this idea will be extensively applied to railways and also to ocean liners, though in the latter case the conditions are not quite so favorable. How the railroad companies can persist in using the ordinary locomotive is a mystery. By providing an engine generating electricity and operating with the current motors under the cars a train can be propelled with greater speed and more economically. In France this has already been done by Heilman, and although his machinery was not the best, the results he obtained were creditable and encouraging. I have calculated that a notable gain in speed and economy can also be secured in ocean liners, on which the improvement is particularly desirable for many reasons. It is very likely that in the near future oil will be adopted as fuel, and that will make the new method of propulsion all the more commendable. The electric manufacturing companies will scarcely be able to meet this new demand for generators and motors.

In automobiles practically nothing has been done in this direction, and yet it would seem they offer the greatest opportunities for application of this principle. The question, however, is which motor to employ - the direct-current or my induction motor. The former has certain preferences as regards the starting and regulation, but the commutators and brushes are very objectionable on an automobile. In view of this I would advocate the use of the induction motor as an ideally simple machine which can never get out of order. The conditions are excellent, inasmuch as a very low frequency is practicable and more than three phases can be used. The regulation should offer little difficulty, and once an automobile on this novel plan is produced its advantages will be readily appreciated.

Yours very truly,
N. Tesla

Tesla on Subway Dangers
New York Sun, June 16, 1905

The flooding of the subway is a calamity apt to repeat itself. As your readers will remember, it did not occur for the first time last Sunday. Water, like fire, will break loose occasionally in spite of precautions. It will never be possible to guard against a casual bursting of a main; for while the conduit can be safely relied upon under normal working conditions, any accidental obstruction to the flow may cause a pressure which no pipe or joint can withstand.

In fact, if we are to place faith in the gloomy forecasts of Commissioner Oakley, who ought to know, such floods may be expected to happen frequently in the future. In view of this it seems timely to call to public attention a danger inherent to the electrical equipment which has been thrust upon the Interborough Company by incompetent advisers.

The subway is bound to be successful, and would be so if the cars were drawn by mules, for it is the ideal means of transportation in crowded cities. But the full measure of success of which it is capable will be attained only when the financiers shall say to the electric companies: "Give us the best, regardless of expense."

It is to be regretted that this important pioneering enterprise, in other respects ably managed and engineered, should have been treated with such gross neglect in its most vital feature. No opportunity was given to myself, the inventor and patentee of the system adopted in the subway and the elevated roads, for offering some useful suggestion, nor was a single electrician or engineer of the General Electric and Westinghouse companies consulted, the very men who should have been thought of first of all.

Once large sums of money are invested in a defective scheme it is difficult to make a change, however desirable it may be. The movement of new capital is largely determined by previous investment. Even the new roads now planned are likely to be equipped with the same claptrap devices, and so the evil will grow. *"Das eben ist der Fluch der boesen Thut, das sie fortzeugend Boeses muss gebaeren."*

The danger to which I refer lies in the possibility of generating an explosive mixture by electrolytic decomposition and thermic dissociation of the water through the direct currents used in the operation of the cars. Such a process might go on for hours and days without being noticed; and with currents of this kind it is scarcely practicable to avoid it altogether.

It will be recalled that an expert found the percentage of free oxygen in the subway appreciably above that which might reasonably have been expected in such a more or less stagnated channel. I have never doubted the correctness of that analysis and have assumed that oxygen is being continuously set free by stray currents passing through the moist ground. The total amperage of the normal working current in the tunnel is very great, and in case of flooding would be sufficient to generate not far from 100 cubic feet of hydrogen per minute. Inasmuch, however, as in railway operation the fuses must be set hard, in order to avoid frequent interruption of the service by their blowing out, in such an emergency the current would be of . much greater volume and hydrogen would be more abundantly liberated.

It is a peculiar property of this gas that it is capable of exploding when mixed with a comparatively large volume of air, and any engineer can convince himself by a simple calculation that, say, 100,000 cubic feet of explosive might be formed before the danger is discovered, reported and preventive measures taken. What the effect of such an explosion might be on life and property is not pleasant to contemplate. True, such a disaster is not probable, but the present electrical equipment makes it possible, and this possibility should be, by all means, removed.

The oppressiveness of the tunnel atmosphere is in a large measure due to the heat supplied by the currents, and to the production of nitrous acid in the arcs, which is enhanced by rarefaction of the air through rapid motion. Some provision for ventilation is imperative. But ventilation will not do away with the danger I have pointed out. It can be completely avoided only by discarding the direct current.

I should say that the city authorities, for this if for no other reason, should forbid its use by a proper act of legislation. Meanwhile, the owners of adjacent property should object to its employment, and the insurance companies should refuse the grant of policies on such property except on terms which it may please them to make.

N. Tesla

Tesla's Reply to Edison
English Mechanic and World of Science — July 14, 1905

As we said last week, Mr. Edison was reported to have said in an interview of the *New York World* that he did not believe Tesla would be able to talk round the world, but that he thought Marconi would, sooner or later, perfect his system.

Nikola Tesla has replied. He says:
In the course of certain investigations which I carried on for the purpose of studying the effects of lightning discharges upon the electrical condition of the earth I observed that sensitive receiving instruments arranged so as to be capable of responding to electrical disturbances created by the discharges at times failed to respond when they should have done so, and upon inquiring into the causes of this unexpected behavior I discovered it to be due to the character of the electrical waves which were produced in the earth by the lightning discharges, and which had nodal regions following at definite distances the shifting source of the disturbances. From data obtained in a large number of observations of the maxima and minima of these waves I found their length to vary approximately from twenty-five to seventy kilometers, and these results and theoretical deductions led me to the conclusion that waves of this kind may be propagated in all directions over the globe, and that they may be of still more widely differing lengths, the extreme limits being imposed by the physical dimensions and properties of the earth. Recognizing in the existence of these waves an unmistakable evidence that the disturbances created had been conducted from their origin to the most remote portions of the globe, and had been thence reflected, I conceived the idea of producing such waves in the earth by artificial means, with the object of using them for many useful purposes for which they are or might be found applicable.

Beat Lightning Flashes
This problem was rendered extremely difficult, owing to the immense dimensions of the planet, and consequently enormous movement of electricity or rate at which electrical energy had to be delivered in order to approximate, even in a remote degree, movements or rates which are manifestly attained in the displays of electrical forces in nature, and which seemed at first unrealizable by any human agencies; but by gradual and

continuous improvements of a generator of electrical oscillations, which I have described in my Patents Nos. 645,576 and 649,621, I finally succeeded in reaching electrical movements or rates of delivery of electrical energy not only approximately, but, as shown in comparative tests and measurements, actually surpassing those of lightning discharges and by means of this apparatus I have found it possible to reproduce, whenever desired, phenomena in the earth the same as or similar to those due to such discharges. With the knowledge of the phenomena discovered by me, and the means at command for accomplishing these results, I am enabled, not only to carry out many operations by the use of known instruments, but also to offer a solution for many important problems involving the operation or control of remote devices which, for want of this knowledge and the absence of these means, have heretofore been entirely impossible. For example, by the use of such a generator of stationary waves and receiving apparatus properly placed and adjusted in any other locality, however remote, it is practicable to transmit intelligible signals, or to control or actuate at will any one or all of such apparatus for many other important and valuable purposes, as for indicating whenever desired the correct time of an observatory, or for ascertaining the relative position of a body or distance of the same with reference to the given point, or for determining the course of a moving object, such as a vessel at sea, the distance traversed by the same or its speed; or for producing many other useful effects at a distance dependent on the intensity, wavelength, direction or velocity of movements, or other feature or property of disturbances of this character.

A Bit of Sarcasm

Permit me to say on this occasion that if there exist to-day no facilities for wireless telegraphic and telephone communication between the most distant countries, it is merely because a series of misfortunes and obstacles have delayed the consummation of my labors, which might have been completed three years ago. In this connection I shall well remember the efforts of some, unwise enough to believe that they can gain an advantage by throwing sand in the eyes of the people and retarding the progress of invention. Should the first messages across the seas prove calamitous to them, it will be a punishment regrettable but fully deserved.

Nikola Tesla

Tesla on the Peary North Pole Expedition
New York Sun — July 16, 1905

To the Editor of the New York Sun:

Everybody must have been pleased to learn that Commodore Peary has finally obtained the financial assistance which will enable him to start without further delay on his important journey. Let us wish the bold navigator the most complete success in his perilous undertaking, in the interest of humanity as well as for his own and his companions' sake and the gratification of the generous donors who have aided him. But, while voicing these sentiments, let us hope that Peary's will be the last attempt to reach the pole in this slow, penible and hazardous way.

We have already sufficiently advanced in the knowledge of electricity and its applications to avail ourselves of better means of transportation, enabling us to reach and to explore without difficulty and in a more perfect manner not only the North, but also the South Pole, and any other still unknown regions of the earth's surface. I refer to the facilities afforded in this respect by the transmission of electrical energy without wires and aerial navigation, which has found in the novel art its ideal solution.

Many of your readers will, no doubt, be under the impression that I am speaking merely of possibilities. As a matter of fact, from the principles involved and the experiments which I have actually performed, not only is the practical success of such distribution of power reduced to a degree of mathematical certitude, but the transmission can be-effected with an economy much greater than possible by the present method involving the use of wires.

It would not take long to build a plant for purposes of aerial navigation and geographical research, nor would it cost as much as might be supposed. Its location would be perfectly immaterial. It might be at the Niagara, or at the Victorian Falls in Africa, without any appreciable difference in the power collected in a flying machine or other apparatus.

A popular error, which I have often opportunity to correct, is to believe that the energy of such a plant would dissipate itself in all directions. This is not so, as I have pointed out in my technical publications. Electricity is displaced by the transmitter in all directions, equally through the earth and the air; that is true, but energy is expended only at the place where it is collected and used to perform some work. To illustrate, a plant of 10,000 hp, such as I have been planning, might be running full blast at Niagara, and there might be but one flying machine, of, say, 50 hp operating in some distant place, the location being of absolutely no consequence. In this case 50 hp would be all the power furnished by the plant to the rest of the universe. Although the electrical oscillations would manifest themselves all over the earth, at the surface as well as high in the air, virtually no power would be consumed. My experiments have shown that the entire electrical movement which keeps the whole globe a-tremble can be maintained with but a few horsepower. Apart from the transmitting and receiving apparatus, the only loss incurred is the energy radiated in the form of Hertzian or electro-magnetic waves, which can be reduced to any entirely insignificant quantity.

I appreciate the difficulty which your non-technical readers must experience in comprehending the working of this system. To gain a rough idea, let them imagine the transmitter and the earth to be two elastic bags, one very small and the other immense, both being connected by a tube and filled with some incompressible fluid. A pump is provided for forcing the fluid from one into the other, alternately and in rapid succession. Now, to produce a great movement of the fluid in a bag of such enormous size as the earth would require a pump so large that it would be a greater task to construct it than to build a thousand Egyptian pyramids. But there is a way of accomplishing this with a pump of very small dimensions. The bag connected to the earth is elastic, and when suddenly struck vibrates at a certain rate. The first artifice consists in so designing and adjusting the parts that the natural vibrations of the bag are in synchronism with the strokes of the pump. Under such conditions the bag is set into violent vibrations, and the fluid is made to rush in and out with terrific force. But the immense bag - the earth, is still comparatively undisturbed. Its size, however, does not exempt it from the laws of nature, and just as the small bag, so too the earth, responds to certain impulses. This fact I discovered in 1899.

The second artifice is to so adjust the transmitter that it will furnish these particular impulses. When all is properly done the large bag is thrown into spasms of vibration, and the effects are bewildering. But no power is yet transmitted, and all this colossal movement requires little energy to maintain. It is like an engine running without load.

Next let your readers imagine that at any place where it may be desired to deliver energy a small elastic bag, not unlike the first, is connected to the large one through a tube. The third artifice consists in so proportioning the parts that the attachment will be responsive to the impulse transmitted, this resulting in a great intensification of the vibration of the bag. Still the pump will not furnish power until these vibrations are made to do work of some kind.

To conduce to an understanding of the fourth artifice, that of "individualization," let your readers follow me a step further, and conceive the flow of energy to any point can be controlled from the place where the pump is located at will, and with equal facility and precision, regardless of distance, and, furthermore, through a device such as the combination lock of a safe, they will then have a crude idea of the processes involved. But only when they realize that all these and many other processes not mentioned, and related to one another like the links of a chain, are completed in a fraction of a second, will your readers be able to appreciate the magical potencies of electrical vibrations and form a conception of the miracles which a skilled electrician can perform by the use of these appliances.

I earnestly hope that in the near future the conditions will be favorable for the construction of a plant such as I have proposed. As soon as this is done it will be possible to adapt electrical motors to flying machines of the type popularized by Santos Dumont. There will be no necessity of carrying a generator or store of motive energy and consequently the machine will be much lighter and smaller. Owing to this and also to the greater power available for propulsion, the speed will be considerably increased. But a few of such machines, properly equipped with photographic and other appliances, will be sufficient to give us in a short time an exact knowledge of the entire earth's surface. It should be borne in mind, however, that for the ordinary uses of a single person a very small machine of not more than one-quarter horse-power, corresponding to the work of two men, would be amply sufficient so that when the first plant of 10,000 hp is installed, the commodity of aerial flight can be offered to a great many individuals all the world over. I can conceive of no improvement which would be more efficient in the furtherance of civilization than this.

N. Tesla

Signaling to Mars —A Problem of Electrical Engineering

Harvard Illustrated — March, 1907

In the early part of 1900, still vividly impressed by certain observations, I had made shortly before, and feeling that the time had come to prepare the world for an experiment which will soon be undertaken, I dwelt on the practicability of interplanetary signaling in an article which appeared in the June number of *Century Magazine* of the same year. In order to correct an erroneous report which gained wide circulation, a statement was published in *Collier's Weekly* of Feb. 9, 1901, defining my position in general terms. Ever since, my thoughts have been centered on the subject, and my original conviction has been strengthened both by reflection and suggestion.

Chief among the stimulating influences was the revelatory work of Percival Lowell, described in a volume with which the observatory, bearing his name, has honored me. No one can look at his globe of Mars without a feeling of profound astonishment, if not awe. These markings, still imperfectly discerned and incomprehensible, but evidently intended for a useful purpose, may they not contain a record of deep meaning left by a superior race, perhaps extinct, to tell its young brethren in other worlds of secrets discovered, of life and struggle, of their own terrible fate? What mighty pathos and love in such a gigantic drama of the universe: But let us hope that the astronomer has seen true, that Mars is not a cold grave, but the abode of happy intelligent creatures, from whom we may learn. In the light of this glorious possibility, signaling to that planet presents itself as a preeminently practical proposition which, to carry out, no human sacrifice could be too great. Can it be done? What chance is there that it will be done?

These questions will be answered definitely the moment all doubt as to the existence of highly developed beings on Mars is dispelled. The straightness of the lines on Lowell's map, their uniform width and other geometrical peculiarities, do not, themselves, appeal to me as strong proofs of artificiality. I should think that a planet large enough not to be frozen stiff in a spasm of volcanic action, like our moon, must, in the course of eons, have all its mountains leveled, the valleys filled, the rocks ground to sand, and ultimately assume the form of a smooth spheroid, with all its rivers flowing in geodetically straight lines. The uniform width of the waterways can be consistently explained, their crossings, however odd and puzzling, might be accidental. But I quite agree with Professor Morse, that this whole wonderful map produces the absolute and irresistible conviction, that these "canals" owe their existence to a guiding intelligence. Their great size is not a valid argument to the contrary. It would merely imply that the Martians have harnessed the energy of waterfalls. We know of no other source of power competent to explain such tremendous feats of engineering. They could not be accomplished by capturing the sun's rays or abstracting heat imparted to the atmosphere, for this, according to our best knowledge, would require clumsy and inefficient machinery. Large falls could be obtained near the polar caps by extensive dams. While much less effective than our own, they could well furnish several billions of horse-power. It should be borne in mind that many Martian tasks in mechanical engineering are much easier than the terrestrial, on account of the smaller mass of the planet and lesser density, which, in the superficial layers, may be considerably below the mean. To a still greater degree this is true of electrical engineering. Taking into account the space encompassed by Mars, a system of wireless transmission of energy, such as I have perfected, would be there much more advantageously applied, for, under similar conditions, a receiving circuit would collect sixteen times as much energy as on the earth.

The astonishing evidences furnished by Lowell are not only indicative of organic life, but they make it appear very probable that Mars is still populated; and furthermore, that its inhabitants are highly developed intelligent beings. Is there any other proof of such existence? I answer, emphatically, yes, prompted both by an instinct which has never yet deceived me, and observation. I refer to the strange electrical disturbances, the discovery of which I announced six years ago. At that time I was only certain that they were of planetary origin. Now, after mature thought and study, I have come to the positive conclusion that they must emanate from Mars.

Life, as a great philosopher has said, is but a continuous adjustment to the environment. Similar conditions must bring forth similar automata. We can have no idea what a Martian might be like, but he certainly has sensitive organs, much as our own, responsive to external stimuli. The indications of these instruments must be real and true. A straight line, a geometrical figure, a number, must convey to his mind a clear and definite conception. He ought to think and reason like ourselves. If he breathes, eats and drinks, he is moved by motives and desires not very different from our own. Such colossal transformation as is observable on the face of Mars could not have been wrought except by beings ages ahead of us in development. What wonder, then, if they have maps of this, our globe, as perfect as Professor Pickering's photographs of the moon? What wonder if they are signaling to us? We are sufficiently advanced in electrical science to know that their task is much easier than ours. The question is, can we transmit electrical energy to that immense distance? This I think myself competent to answer in the affirmative.

N. Tesla

Tuned Lightning

English Mechanic and World of Science, March 8, 1907

I read with interest an article in the Sunday World of Jan. 20 on "Tuned Lightning," described as a mysterious new energy, which is to turn every wheel on earth, and is supposed to have been recently discovered by the Danish inventors Waldemar Poulsen and P. O. Pederson.

From other reports I have gathered that these gentlemen have so far confined themselves to the peaceful production of miniature bolts not many inches long, and I am wondering what an account of their prospective achievements would read like if they had succeeded in obtaining, like myself, electrical discharges of 100 ft., far surpassing lightning in some features of intensity and power.

In view of their limited Jovian experience, the program outlined by the Danish engineers is rather extensive, Lord Armstrong's vast resources notwithstanding. Naturally enough, I shall look with interest to their telephoning across the Atlantic, supplying light and propelling airships without wires. *Anch in suito pittore.* (I, too, am a painter.) In the mean time it may not be amiss to state here incidentally that all the essential processes of and appliances for the generation, transmission, transformation, distribution, storage, regulation, control, and economic utilization of "tuned lightning" have been patented by me, and that I have long since undertaken, and am sparing no effort to render these advances instrumental in insuring the welfare, comfort, and convenience, primarily, of my fellow citizens.

There is nothing remarkable in the demonstration reported to have been made before Sir William Preece and Prof. Sylvanus P. Thompson, nor is there any novelty in the electrical devices employed. The lighting of arc lamps through the human body, the fusing of a piece of copper in mid-air, as described, are simple experiments which by the use of my high-frequency transformers any student of electricity can readily perform. They teach nothing new, and have no bearing on wireless transmission, for the actions virtually cease at a distance of a few feet from the source of vibratory energy. Years ago I gave exhibitions of similar and other much more striking experiments with the same kind of apparatus, many of which have been illustrated and explained in technical journals. The published records are open to inspection.

Regardless of all that, the Danish inventors have not as yet offered the slightest proof that their expectations are realizable, and before advancing seriously the claim that an efficient wireless distribution of light and power to great distances is possible, they should, at least, repeat those of my experiments which have furnished this evidence.

A scientific audience cannot help being impressed by a display of interesting phenomena, but the originality and significance of a demonstration such as that referred to can only be judged by an expert possessed of full knowledge and capable of drawing correct conclusions. A novel effect, spectacular and surprising, might be quite unimportant, while another, seemingly trifling, is of the greatest consequence.

To illustrate, let me mention here two widely different experiments of mine. In one the body of a person was subjected to the rapidly-alternating pressure of an electrical oscillator of two and a half million volts; in the other a small incandescent lamp was lighted by means of a resonant circuit grounded on one end, all the energy being drawn through the earth electrified from a distant transmitter.

The first presents a sight marvelous and unforgettable. One sees the experimenter standing on a big sheet of fierce, blinding flame, his whole body enveloped in a mass of phosphorescent wriggling streamers like the tentacles of an octopus. Bundles of light stick out from his spine. As he stretches out the arms, thus forcing the electric fluid outwardly, roaring tongues of fire leap from his fingertips. Objects in his vicinity bristle with rays, emit musical notes, glow, grow hot. He is the center of still more curious actions, which are invisible. At each throb of the electric force myriads of minute projectiles are shot off from him with such velocities as to pass through the adjoining walls. He is in turn being violently bombarded by the surrounding air and dust. He experiences sensations which are indescribable.

A layman, after witnessing this stupendous and incredible spectacle, will think little of the second modest exhibit. But the expert will not be deceived. He realizes at once that the second experiment is ever so much more difficult to perform and immensely more consequential. He knows that to make the little filament glow, the entire surface of the planet, two hundred million square miles, must be strongly electrified. This calls for peculiar electrical activities, hundreds of times greater than those involved in the lighting of an arc lamp through the human body. What impresses him most, however, is the knowledge that the little lamp will spring into the same brilliancy anywhere on the globe, there being no appreciable diminution of the effect with the increase of distance from the transmitter.

This is a fact of overwhelming importance, pointing with certitude to the final and lasting solution of all the great social, industrial, financial, philanthropic, international, and other problems confronting humanity, a solution of which will be brought about by the complete annihilation of distance in the conveyance of intelligence, transport of bodies and materials, and the transmission of the energy necessary to man's existence. More light has been thrown on this scientific truth lately through Prof. Slaby's splendid and path-breaking experiment in establishing perfect wireless telephone connection between Naum and Berlin, Germany, a distance of twenty miles. With apparatus properly organised such telephonic communication can be effected with the same facility and precision at the greatest terrestrial distance.

The discovery of the stationary terrestrial waves, showing that, despite its vast extent, the entire planet can be thrown into resonant vibration like a little tuning fork; that electrical oscillations suited to its physical properties and dimensions pass through it unimpeded, in strict obedience to a simple mathematical law, has proved beyond the shadow of a doubt that the earth, considered as a channel for conveying electrical energy, even in such delicate and complex transmissions as human speech or musical composition, is infinitely superior to a wire or cable, however well designed.

Very soon it will be possible to talk across an ocean as clearly and distinctly as across a table. The first practical success, already forecast by Slaby's convincing demonstration, will be the signal for revolutionary improvements which will take the world by storm.

However great the success of the telephone, it is just beginning its evidence of usefulness. Wireless transmission of speech will not only provide new but also enormously extend existing facilities. This will be merely the forerunner of ever so much more important development, which will proceed at a furious pace until, by the application of these same great principles, the power of waterfalls can be focused whenever desired; until the air is conquered, the soil fructified and embellished; until, in all departments of human life distance has lost its meaning, and even the immense gulf separating us from other worlds is bridged.

Nikola Tesla

Tesla's Wireless Torpedo

New York Times — March 20, 1907

To the Editor of The *New York Times*:

A report in the *Times* of this Morning says that I have attained no practical results with my dirigible wireless torpedo. This statement should be qualified. I have constructed such machines, and shown them in operation on frequent occasions. They have worked perfectly, and everybody who saw them was amazed at their performance.

It is true that my efforts to have this novel means for attack and defense adopted by our Government have been unsuccessful, but this is no discredit to my invention. I have spent years in fruitless endeavor before the world recognized the value of my rotating field discoveries which are now universally applied. The time is not yet ripe for the telautomatic art. If its possibilities were appreciated the nations would not be building large battleships. Such a floating fortress may be safe against an ordinary torpedo, but would be helpless in a battle with a machine which carries twenty tons of explosive, moves swiftly under water, and is controlled with precision by an operator beyond the range of the largest gun.

As to projecting wave-energy to any particular region of the globe, I have given a clear description of the means in technical publications. Not only can this be done by the use of my devices, but the spot at which the desired effect is to be produced can be calculated very closely, assuming the accepted terrestrial measurements to be correct. This, of course, is not the case. Up to this day we do not know a diameter of the globe within one thousand feet. My wireless plant will enable me to determine it within fifty feet or less, when it will be possible to rectify many geodetical data and make such calculations as those referred to with greater accuracy.

Nikola Tesla

New York, March 19, 1907

Wireless on Railroads

New York Times — March 26, 1907

To the Editor of The New York Times:
No argument is needed to show that the railroads offer opportunities for advantageous use of a practical wireless system. Without question, its widest field of application is the conveyance to the trains of such general information as is indispensable for keeping the traveler in touch with the world. In the near future a telegraphic printer of news, a stock ticker, a telephone, and other kindred appliances will form parts of the regular wireless equipment of a railroad train. Success in this sphere is all the more certain, as the new is not antagonistic, but, on the contrary, very helpful to the old. The technical difficulties are minimized by the employment of a transmitter the effectiveness of which is unimpaired by distance.

In view of the great losses of life and property, improved safety devices on the cars are urgently needed. But upon careful investigation it will be found that the outlook in this direction is not very promising for the wireless art. In the first place the railroads are rapidly changing to electric motive power, and in all such cases the lines become available for the operation of all sorts of signaling apparatus, of which she telephone is by far the most important. This valuable improvement is due to Prof. J. Paley, who introduced it in Germany eight years ago. By enabling the engineer or conductor of any train to call up any other train or station along the track and obtain full and unmistakable information, the liability of collisions and other accidents will be greatly reduced. Public opinion should compel the immediate adoption of this invention.

Those roads which do not contemplate this transformation might avail themselves of wireless transmission for similar purposes, but inasmuch as every train will require in addition to a complete outfit an expert operator, many roads may prefer to use a wire, unless a wireless telephone can be offered to them.

Nikola Tesla
New York, March 25, 1907

Nikola Tesla Objects

New York Times — May 2, 1907

To the Editor of The New York Times:

I have been much surprised to read in The Times of Sunday, April 21, that Admiral H. N. Manney, U.S.N. attributes a well-known invention of mine, a process for the production of continuous electrical oscillations by means of the electric arc and condenser, to Valentine Poulsen, the Danish engineer. This improvement has been embodied by me in numerous forms of apparatus identified with my name; and I have described it minutely in patents and scientific articles. To quote but one of many references, I may mention my experimental lecture on "Light and Other High-Frequency Phenomena," published under the auspices of the Franklin Association, for which both of these societies have distinguished me.

I share with Admiral Manney in the gratification that we are in the lead, and particularly that wireless messages have been transmitted from Pensacola to Point Lorne. Inasmuch, however, as this feat could not have been accomplished except by the use of some of my own devices, it would have been a graceful act on his part to bring this feat to the attention of the wireless conference. My theory has always been that military men are superior to civilians in courtesy. I have not been discouraged by the refusal of our Government to adopt my wireless system six years ago, when I offered it, not by the unpleasant prospect of my passing through the experiences described by Mark Twain in his story of the beef contract, but *I* see no reason why I should be deprived of a well-earned honor and satisfaction.

The Times has hurt me grievously; not by accusing me of commercialism, nor by its unkind editorial comments on those letters I wrote, in condemnation of my system of power transmission in the Subway. It is another injury, perhaps, unintentional, which I have felt most keenly.

The editor of The Times may not have known that I am a student of applied mathematics when he permitted a fellow student of mine to insinuate in The Times of March 28 that I avail myself of inventions of others. I cannot permit such ideas to gain ground in this community, and, just to illuminate the situation, I shall quote from the leading electrical paper, *The London*

Electrician, referring to some wireless plants of Braun and Marconi: "The spark occurs between balls in the primary circuit of a Tesla coil. The air wire is in series with a Tesla transformer. The generating plant is virtually a Poldhu in miniature. Evidently Braun, like Marconi, has been converted to the high-potential methods introduced by Tesla." Needless to add that this substitution of the old, ineffective Hertzian appliances for my own has not been authorized by me.

My fellow-student can rest assured that I am scrupulously respecting the rights of others. If I were not prompted to do so by a sense of fairness and pride I would be by the power I have of inventing anything I please.

N. Tesla

New York, April 30, 1907

Mr. Tesla on the Wireless Transmission of Power

N. Y. World May 19, 1907

To the Editor of The World:

I have enjoyed very much the odd prediction of Sir Hugh Bell, President of the Iron and Steel Institute, with reference to the wireless transmission of power, reported in The World of the 10th inst.

With all the respect due to that great institution I would take the liberty to remark that if its President is a genuine prophet he must have overslept himself a trifle. Sir Hugh would honor me if he would carefully peruse my British patent No. 8,200, in which I have recorded some of my discoveries and experiments, and which may influence him to considerably reduce his conservative estimate of one hundred years for the fulfillment of his prophecy.

Personally, basing myself on the knowledge of this art to which I have devoted my best energies, I do not hesitate to state here for future reference and as a test of accuracy of my scientific forecast that flying machines and ships propelled by electricity transmitted without wire will have ceased to be a wonder in ten years from now. I would say five were it not that there is such a thing as "inertia of human opinion" resisting revolutionary ideas.

It is idle to believe that because man is endowed with higher attributes his material evolution is governed by other than general physical laws. If the genius of invention were to reveal to-morrow the secret of immortality, of eternal beauty and youth, for which all humanity is aching, the same inexorable agents which prevent a mass from changing suddenly its velocity would likewise resist the force of the new knowledge until time gradually modifies human thought.

What has amused me still more, however, is the curious interview with Lewis Nixon, the naval contractor, printed in the World of the 11th inst. Is it possible that the famous designer of the Oregon is not better versed in editorial matters than some of my farming neighbors of Shoreham? One cannot escape that conviction.

We are not in the dark as regards the electrical energy contained in the earth. It is altogether too insignificant for any industrial use. The current circulating through the globe is of enormous volume but of small tension, and could perform but little work.

Beside, how does Nixon propose to coax the current from the natural path of low resistance into an artificial channel of high resistance? Surely he knows that water does not flow up hill. It is absurd of him to compare the inexhaustible dynamic energy of wind with the magnetic energy of the earth, which is minute in amount and in a static condition.

The torpedo he proposes to build is not novel. The principle is old. I could refer him to some of my own suggestions of nine years ago. There are many practical difficulties in the carrying out of the idea, and as much better means for destroying a submarine are available it is doubtful that such a torpedo will ever be constructed.

Nixon has failed to grasp that in my wireless system the effect does not diminish with distance. The Hertz waves have nothing to do with it except that some of my apparatus may be used in their production. So too a Kohinoor might be employed to cut window-glass. And yet, the seeming paradox can be easily understood by any man of ordinary intelligence.

Imagine only that the earth were a hollow shell or reservoir in which the transmitter would compress some fluid, as air, for operating machinery in various localities. What difference would it make when this reservoir is tapped to supply the compressed fluid to the motor? None whatever, for the pressure is the same everywhere. This is also true of my electrical system, with all considerations in its favor. In such a mechanical system of power distribution great losses are unavoidable and definite limits in the quality of the energy transmitted exist. Not so in the electrical wireless supply. It would not be difficult to convey to one of our liners, say, 50,000 horsepower from a plant located at Niagara, Victoria or other waterfall, absolutely irrespective of location. In fact, there would not be a difference of more than a small fraction of one per cent, whether the source of energy be in the vicinity of the vessel or 12,000 miles away, at the antipodes.

Nikola Tesla
New York, May 16, 1907.

Can Bridge the Gap to Mars

New York Times — June 23, 1907

To the Editor of the New York Times:
You have called me an "inventor of some useful pieces of electrical apparatus." It is not quite up to my aspirations, but I must resign myself to my prosaic fate. I cannot deny that you are right.

Nearly four million horse power of waterfalls are harnessed by my alternating current system of transmission, which is like saying that one hundred million men untiring, consuming nothing, receiving no pay - are laboring to provide for one hundred million tons of coal annually. In this great city the elevated roads, the subways, the Street railways are operated by my system, and the lamps and other electrical appliances get the current through machinery of my invention. And as in New York so all the world over where electricity is introduced. The telephone and incandescent lamp fill specific and minor demands, electric power meets the many general and sterner necessities of life. Yes, I must admit, however reluctantly, the truth of your unflattering contention.

But the greater commercial importance of this invention of mine is not the only advantage I have over my celebrated predecessors in the realm of the useful, who have given us the telephone and the incandescent lamp. Permit me to remind you that I did not have, like Bell, such powerful help as the Reis telephone, which reproduced music and only needed a deft turn of an adjusting screw to repeat the human voice; or such vigorous assistance as Edison found in the incandescent lamps of King and Starr, which only needed to be made of high resistance. Not at all. I had to cut the path myself, and my hands are still sore. All the army of my opponents and detractors was ever able to drum up against me in a fanatic contest has simmered down to a short article by an Italian - Prof. Ferraris - dealing with an abstract and meaningless idea of a rotating magnetic pole and published years after my discovery, months even after my complete disclosure of the whole practically developed system in all its essential universally adopted features. It is a publication, pessimistic and discouraging, devoid of the discoverer's virility and force, devoid of results, utterly wanting in the faith and devotion of the inventor, a defective and belated record of a good but feeble man whose only response to my whole-souled brother greeting was a plaintive cry of priority - a sad contrast to the strong and equanimous Schallenberger, a true American engineer, who stoically bore the pain that killed him.

A fundamental discovery or original invention is always useful, but it is often more than that. There are physicists and philosophers to whom the marvelous manifestations of my rotating magnetic field, the suggestive phenomena of rotation without visible connection, the ideal beauty of my induction motor with its contactless armature, mean quite as much as the thousands of millions of dollars invested in enterprises of which it is the foundation.

And this is true of all my discoveries, inventions, and scientific results which I have since announced, for I have never invented what immediate necessity suggested, but what I found as most desirable to invent, irrespective of time. Let me tell you only of one - my "magnifying transmitter," a machine with which I have passed a current of one hundred amperes around the globe, with which I can make the whole earth loudly repeat a word spoken in the telephone, with which I can easily bridge the gulf which separates us from Mars. Do you mean to say that my transmitter is nothing more than a "useful piece of electrical apparatus"?

I do not wish to enlarge on this for obvious reasons. To be compelled by taciturn admirers to dwell on my own achievements is hurting my delicate sensibilities, but as I observe your heroic and increasing efforts in praising your paper, while your distinguished confreres maintain on its merits a stolid silence, I feel that there is, at least, one man in New York able to appreciate the incongeniality of the correspondence. Allow me to ask you just one or two questions in regard to a work which I began in 1892, inspired by a high tribute from Lord Rayleigh at the Royal Institution, most difficult labor which I have carried on for years, encouraged by the sympathetic interest and approval of Hemholtz, Lord Kelvin, and my great friends, Sir William Crookes and Sir James Dewar, ridiculed by small men whose names I have seen displayed in vulgar and deceptive advertisements. I refer to my system of wireless transmission of energy.

The principles which it involves are eternal. We are on a conducting body, insulated in space, of definite and unchangeable dimensions and properties. It will never be possible to transmit electrical energy economically through this body and its environment except by essentially the same means and methods which I have discovered, and the system is so perfect now that it admits of but little improvement. Since I have accepted as true your opinion, which I hope will not be shared by posterity, would you mind telling a reason why this advance should not stand worthily beside the discoveries of Copernicus? Will you state why it should not be ever so much more important and valuable to the progress and welfare of man?

We could still believe in the geocentric theory and yet advance virtually as we do. The work of the astronomer would suffer, for some of his deductions would rest on erroneous assumptions. But, after all, we shall never know the intimate nature of things. So long as our perceptions are accurate our logic will be true. No one can estimate to what an extent the great knowledge he conveyed has been instrumental in developing the power of our minds and furthering discovery and invention. Yet, it has left all the pressing material problems confronting us unsolved.

Now my wireless system offers practical solutions for all. The aerial navigation, which now agitates the minds, is only one of its many and obvious applications equally consequential. The waterfalls of this country alone, its greatest wealth, are adequate to satisfy the wants of humanity for thousands of years to come. Their energy can be used with the same facility to dig the Panama Canal as to operate the Siberian Railway or to irrigate and fertilize the Sahara. The Anglo-Saxon race has a great past and present, but its real greatness is in the future, when the water power it owns or controls shall supply the necessities of the entire world.

As to universal peace - if there is nothing in the order of nature which makes war indispensable to the safe and sane progress of man, if that utopian existence is at all possible, it can be only attained through this very means, for all international friction can be traced to but one cause - the immense extension of the planet. My system of wireless transmission completely annihilates distance in all departments of human activity.

If this does not appeal to you sufficiently to recognize in me a discoverer of principles, do me, at least, the justice of calling me an "inventor of some beautiful pieces of electrical apparatus."

Nikola Tesla
New York, June 21, 1907

Sleep From Electricity
New York Times — Oct. 19, 1907

To the Editor of The New York Times:
I have read with interest the reports in The Times of the 13th and 15th inst. referring to Prof. Leduc's discovery of causing sleep by electric means. While it is possible that he has made a distinct advance there is no novelty in the effect itself.

The narcotic influence of certain periodic currents was long ago discovered by me and has been pointed out in some of my technical publications, among which I may mention a paper on "High Frequency Oscillators for Electro Therapeutic and Other Purposes," read before the American Electro Therapeutic Association, Sept. 13, 1898. I have also shown that human tissues offer little resistance to the electric flow and suggested an absolutely painless method of electrocution by passing the currents through the brain. It is very likely that Prof. Leduc has taken advantage of the same general principles though he applies the currents in a different manner.

In one respect, however, my observations are at variance with those reported. From the special dispatch in The Times of the 13th inst. it would appear that sleep is induced the moment the currents are turned on, and that awakening follows as soon as the electrodes are withdrawn. It is, of course, impossible to tell how strong a current was employed, but the resistance of the head might have been, perhaps, 3,000 ohms, so that at thirty volts the current could have been only about 1-100 of an ampere. Now, I have passed a current of at least 5,000 times stronger through my head and did not lose consciousness, but I invariably fell into a lethargic sleep some time after. This fact impresses me with certain arguments of Prof. Barker of Columbia University in your issue of Sept. 15.

I have always been convinced that electric anaesthesia will become practical, but the application of currents to the brain is so delicate and dangerous an operation that the new method will require long ark careful experimentation before it ,can be used with certitude.
Nikola Tesla
New York, Oct. 16, 1907

Possibilities of "Wireless"

New York Times — Oct. 22, 1907

To the Editor of The New York Times:
In your issue of the 19th inst. Edison makes statements which cannot fail to create erroneous impressions.

There is a vast difference between primitive Hertzwave signaling, practicable to but a few miles, and the great art of wireless transmission of energy, which enables an expert to transmit, to any distance, not only signals, but power in unlimited amounts, and of which the experiments across the Atlantic are a crude application. The plants are quite inefficient, unsuitable for finer work, and totally doomed to an effect less than one percent of that I attained in my test in 1899.

Edison thinks that Sir Hiram Maxim is blowing hot air. The fact is my Long Island plant will transmit almost its entire energy to the antipodes, if desired. As to Martin's communication I can only say, that I shall be able to attain a wave activity of 800,000,000 horse power and a simple calculation will show, that the inhabitants of that planet, if there be any, need not have a Lord Raleigh to detect the disturbance.

Referring to your editorial comment of even date, the question of wireless interference is puzzling only because of its novelty. The underlying principle is old, and it has presented itself for consideration in numerous forms. Just now it appears in the novel aspects of aerial navigation and wireless transmission. Every human effort must of necessity create a disturbance. What difference is there in essence, between the commotion produced by any revolutionary idea or improvement and that of a wireless transmitter? The specter of interference has been conjured by Hertzwave or radio telegraphy in which attunement is absolutely impossible, simply because the effect diminishes rapidly with distance. But to my system of energy transmission, based on the use of impulses not sensibly diminishing with distance, perfect attunement and the higher artifice of individualization are practicable. As ever, the ghost will vanish with the wireless dawn.

Nikola Tesla
New York, Oct. 21, 1907.

Tesla on Wireless
New York Daily Tribune — Oct. 25, 1907

To the Editor of The Tribune:

Sir: In so far as wireless art is concerned there is a vast difference between the great inventor Thomas A. Edison and myself, integrally in my favor. Mr. Edison knows little of the theory and practice of electrical vibrations; I have, in this special field, probably more experience than any of my contemporaries. That you are not as yet able to impart your wisdom by wireless telephone to some subscriber in any other part of the world, however remote, and that the presses of your valuable paper are not operated by wireless power is largely due to your own effort and those of some of your distinguished confreres of this city, and to the efficient assistance you have received from my celebrated colleagues, Thomas A. Edison and Michael Pupin, assistant consulting wireless engineers. But it was all welcome to me. Difficulty develops resource.

The transmission across the Atlantic was not made by any device of Mr. Marconi's, but by my system of wireless transmission of energy, and I have already given notice by cable to my friend Sir James DeWar and the Royal Institution of this fact. I shall also request some eminent man of science to take careful note of the whole apparatus, its mode of operation, dimensions, linear and electrical, all constants and qualitative performance, so as to make possible its exact reproduction and repetition of the experiments. This request is entirely impersonal. I am a citizen of the United States, and I know that the time will come when my busy fellow citizens, too absorbed in commercial pursuits to think of posterity, will honor my memory. A measurement of the time interval taken in the passage of the signal necessary to the full and positive demonstration will show that the current crosses the ocean with a mean speed of 625,000 miles a second.

The Marconi plants are inefficient, and do not lend themselves to the practice of two discoveries of mine, the "art of individualization," that makes the message non-interfering and non-interferable, and the "stationary waves," which annihilate distance absolutely and make the whole earth equivalent to a conductor devoid of resistance. Were it not for this deficiency, the number of words per minute could be increased at will by "individualizing."

You have already commented upon this advance in terms which have caused me no small astonishment, in view of your normal attitude. The underlying principle is to combine a number of vibrations, preferably slightly displaced, to reduce further the danger of interference, active and passive, and to make the operation of the receiver dependent on the co-operative effect of a number of attuned elements. Just to illustrate what can be done, suppose that only four vibrations were isolated on each transmitter. let those on one side be respectively a, b, c, and d. Then the following individualized lines would be ab, ac, ad, bc, bd, cd, abc, abd, acd, bed and abed. The same article on the other side will give similar combinations, and both together twenty-two lines, which can be simultaneously operated. To transmit one thousand words a minute, only forty-six words on each combination are necessary. If the plants were suitable, not ten years, as Edison thinks, but ten hours would be necessary to put this improvement into practice. To do this Marconi would have to construct the plants, and it will then be observed that the indefatigable Italian has departed from universal engineering customs for the fourth time.

Nikola Tesla
New York, Oct. 24, 1907

My Apparatus, Says Tesla
New York Times — Dec. 20, 1907

To the Editor of the New York Times:
I have read with great interest the report in your issue of to-day that the Danish engineer, Waldemar Poulson, the inventor of the interesting device known as the "telegraphone," has succeeded in transmitting accurately wireless telephonic messages over a distance of 240 miles.

I have looked up the description of the apparatus he has employed in the experiment and find that it comprises:

(1) My grounded resonant transmitting circuit;

(2) my inductive exciter;

(3) the so-called "Tesla transformer";

(4) my inductive coils for raising the tension on the condenser;

(5) my entire apparatus for producing undamped or continuous oscillations;

(6) my concatenated tuned transforming circuits;

(7) my grounded resonant receiving transformer;

(8) my secondary receiving transformer.

I note other improvements of mine, but those mentioned will be sufficient to show that Denmark is a land of easy invention.

The claim that transatlantic wireless telephone service will soon be established by these means is a modest one. To my system distance has absolutely no significance. My own wireless plant will transmit speech across the Pacific with the same precision and accuracy as across the table.

Nikola Tesla
New York, Dec. 19, 1907

Nikola Tesla's Forecast for 1908

N. Y. World — Jan. 5, 1908

To the Editor of The World:
Constant and careful study of the state of things in this particular sphere enables an expert to make a forecast fairly accurate of the next state. The seemingly isolated events are to him but links of a chain. As a rule, the signs he notes are so pronounced that he can predict the changes about to take place with certitude. The performance is a mere banality as compared with the piercing view of the inspired into the distant future. This is a forecast - not a prophecy.

The coming year will be great in thought and result. It will mark the end of a number of erroneous ideas which, by their paralyzing effect on the mind, have throttled independent research and hampered progress and development in various departments of science and engineering.

The first to be dispelled is the illusion of the Hertz or electro-magnetic waves. The expert already realizes that practical wireless telegraphy and telephony are possible only by minimizing this wasteful radiation. The results recently attained in this manner with comparatively crude appliances illustrate strikingly the possibilities of the genuine art. Before the close of the year wireless transmission across the Pacific and trans-Atlantic wireless telephony may be expected with perfect confidence. The use of the wireless telephone in isolated districts will spread like fire.

The year will mark the fall of the illusionary idea that action must diminish with distance. By impressing upon the earth certain vibrations to which it responds resonantly, the whole planet is virtually reduced to the size of a little marble, thus enabling the reproduction of any kind of effect, as human speech, music, picture or character whatever, and even the transmission of power in unlimited amounts with exactly the same facility and economy at any distance, however great.

The next twelve months will witness a similar revolution of ideas regarding radio-activity. That there is no such element as radium, polonium or ronium is becoming more and more evident. These are simply deceptive appearances of a modern phlogiston. As I have stated in my early announcement of these emanations before the discovery of Mme. Curie, they are emitted more or less by all bodies, and are all of the same kind - merely effects of shattered molecules, differentiated not by the nature of substance but by size, speed and electrification.

The coming year will dispel another error which has greatly retarded progress of aerial navigation. The aeronaut will soon satisfy himself that an airplane proportioned according to data obtained by Langley is altogether too heavy to soar, and that such a machine, while it will have some uses, can never fly as fast as a dirigible balloon. Once this is fully recognized the expert will concentrate his efforts on the latter type, and before many months are passed it will be a familiar object in the sky.

There is abundant evidence that distinct improvements will be made in ship propulsion. The numerous theories are giving place to the view that what propels the vessel is a reactive jet; hence the propeller is doomed in efficiency at high speed. A new principle will be introduced.

The World is invited to test the accuracy of this forecast at the close of the year.

Nikola Tesla

Mr. Tesla's Vision

New York Times — April 21, 1908

To the Editor of the New York Times:
From a report in your issue of March 11, which escaped my attention, I notice that some remarks I made on the occasion referred to have been misunderstood. Allow me to make a correction..

When I spoke of future warfare I meant that it should be conducted by direct application of electrical waves without the use of aerial engines or other implements of destruction. This means, as I pointed out, would be ideal, for not only would the energy of war require no effort for the maintenance of its potentiality, but it would be productive in times of peace. This is not a dream. Even now wireless power plants could be constructed by which any region of the globe might be rendered uninhabitable without subjecting the population of other parts to serious danger or inconvenience.

What I said in regard to the greatest achievement of the man of science whose mind is bent upon the mastery of the physical universe, was nothing more than what I stated in one of my unpublished addresses, from which I quote: "According to an adopted theory, every ponderable atom is differentiated from a tenuous fluid, filling all space merely by spinning motion, as a whirl of water in a calm lake. By being set in movement this fluid, the ether, becomes gross matter. Its movement arrested, the primary substance reverts to its normal state. It appears, then, possible for man through harnessed energy of the medium and suitable agencies for starting and stopping ether whirls to cause matter to form and disappear, At his command, almost without effort on his part, old worlds would vanish and new ones would spring into being. He could alter the size of this planet, control its seasons, adjust its distance from the sun, guide it on its eternal journey along any path he might choose, through the depths of the universe. He could make planets collide and produce his suns and stars, his heat and light; he could originate life in all its infinite forms. To cause at will the birth and death of matter would be man's grandest deed, which would give him the mastery of physical creation, make him fulfill his ultimate destiny."

Nothing could be further from my thought than to call wireless telephony around the world "the greatest achievement of humanity" as reported. This is a feat which, however stupefying, can be readily performed by any expert. I have myself constructed a plant for this very purpose. The wireless wonders are only seeming, not results of exceptional skill, as popularly believed. The truth is the electrician has been put in possession of a veritable lamp of Aladdin. All he has to do is to rub it. Now, to rub the lamp of Aladdin is no achievement,

If you are desirous of hastening the accomplishment of still greater and further - reaching wonders you can do no better than by emphatically opposing any measure tending to interfere with the free commercial exploitation of water power and the wireless art. So absolutely does human progress depend on the development of these that the smallest impediment, particularly through the legislative bodies of this country, may set back civilization and the cause of peace for centuries.

Nikola Tesla

New York, April 19, 1908

Little Airplane Progress

New York Times — June 8

To the Editor of the New York Times:

It was not a little amusing to read a short time ago how the "great secret" of the airplane was revealed. By surrounding that old device with an atmosphere of mystery one gives life and interest to the report; but the plain fact is that all forms of aerial apparatus are well known to engineers, and can be designed for any specific duty without previous trials and with a fair degree of accuracy. The flying machine has materialized - not through leaps and bounds of invention, but by progress slow and imperceptible, not through original individual effort, but by a combination of the same forces which brought forth the automobile, and the motorboat. It is due to the enterprise of the steel, oil, electrical, and other concerns, who have been instrumental in the improvement of materials of construction and in the production of high-power fuels, as well as to the untiring labors of the army of skilled but unknown mechanics, who have been for years perfecting the internal combustion engine.

There is no salient difference between the dirigible balloon of Renard and Krebs of thirty years ago and that of Santos Oumont with which the bold Brazilian performed his feats. The Langley and Maxim aerodromes, which did not soar, were in my opinion better pieces of mechanism than their very latest imitations. The powerful gasoline motor which has since come into existence is practically the only radical improvement.

So far, however, only the self-propelled machine or aerial automobile is in sight. While the dirigible balloon is rapidly nearing the commercial stage, nothing practical has as yet been achieved with the heavier-than-air machine. Without exception the apparatus is flimsy and unreliable. The motor, too light for its power, gives out after a few minutes run; the propeller blades fly off; the rudder is broken, and, after a series of such familiar mishaps, there comes the inevitable and general smash-up. In strong contrast with these unnecessarily hazardous trials are the serious and dignified efforts of Count Zeppelin, who is building a real flying machine, safe and reliable, to carry a dozen men and provisions over distances of thousands of miles, and with a speed far in excess of those obtained with airplanes.

The limits of improvement in the flying machine, propelled by its own power, whether light or heavy, are already clearly defined. We know very closely what we may expect from the ultimate perfection of the internal combustion engine, the remittances which are to be overcome, and the limitations of the screw propeller. The margin is not very great. For many reasons the wireless transmission of power is the only perfect and lasting solution to reach very high speeds.

In this respect many experts are mistaken. The popular belief that because the air has only one-hundredth the density of water, enormous velocities should be practicable. But it is not so. It should be borne in mind that the air is one hundred times more viscous than water, and because of this alone the speed of the flying machine could not be much in excess of a properly designed aqueous craft.

The airplanes of the Langley type, such as was used by Forman and others with some success, will hardly ever prove a practical aerial machine, because no provision is made for maintaining it in the air in a downward current. This and the perfect balance independently of the navigator's control is absolutely essential to the success of the heavier-than-air machine. These two improvements I am myself endeavoring to embody in a machine of my own design.

Nikola Tesla
New York, June 6, 1908

Tesla on Airplanes
New York Times — Sept. 15

To the Editor of The New York Times:
The chronicler of current events is only too apt to lose sight of the true perspective and real significance of the phases of progress he records. Naturally enough, his opinions on subjects out of the sphere of his special training are frequently defective, but this is inseparable from the very idea of journalism. If an editor were to project himself into the future and view the happenings of the present or of the past in their proper relations he would make a dismal failure of his paper.

The comments upon the latest performances with airplanes afford interesting examples in this respect. What is there so very different between a man flying half an hour and another, using a more powerful machine, an hour, or two, or three? To be sure, in one instance the supporting planes are larger and the gasoline tank bigger, but there is nothing revolutionary in these departures. No one can deny the merit of the accomplishments. The feats are certainly remarkable and of great educational value.

The majority of human beings are unreceptive to novel ideas. The practical demonstrator comes with forceful arguments which enlighten and convince. But they are nothing more than obvious consequences of what has preceded, steps in advance which, taken singly, are of no particular importance, but which, in their totality, make up the conquest of the world by the new idea. If any one stands out more strongly than the other it is merely because it chances to occur at the psychological moment, when incredulity and doubt are giving way to confidence and expectancy. Such work is often brilliant, never great, as some would make believe. To be great it must be original. Of such feature it is absolutely devoid.

Place any of the later airplanes beside that of Langley, their prototype, and you will not find as much as one decided improvement. There are the same old propellers, the same old inclined planes, rudders, and vanes - not a single notable difference. Some have tried to hide their "discoveries." It is like the hiding of an ostrich who buries his head in the sand. Half a dozen aeronauts have been in turn hailed as conquerors and kings of the air. It would have been much more appropriate to greet John D. Rockefeller as such. But for the abundant supply of high-grade fuel we would still have to wait for an engine capable of supporting not only itself but several times its own weight against gravity.

The capabilities of the Langley aerodrome have been most strikingly illustrated. Notwithstanding this, it is not a practical machine. It has a low efficiency of propulsion, and the starting, balancing, and alighting are attended with difficulties. The chief defect, however, is that it is doomed if it should encounter a downward gust of wind. The helicopter is in these respects much preferable, but is objectionable for other reasons. The successful heavier-than-air flier will be based on principles radically novel and will meet all requirements. It will soon materialize, and when it does it will give an impetus to manufacture and commerce such as was never witnessed before, provided only that the Governments do not resort to the methods of the Spanish Inquisition, which have already proved so disastrous to the wireless art, the ideal means for making man absolute master of the air.

Nikola Tesla
New York, Sept. 13, 1908

How to Signal Mars
New York Times — May 23

To the Editor of the New York Times:
Of all the evidence of narrow mindedness and folly, I know of no greater than the stupid belief that this little planet is singled out to be the seat of life, and that all other heavenly bodies are fiery masses or lumps of ice. Most certainly, some planets are not inhabited, but others are, and among these there must exist II life under all conditions and phases of development.

I In the solar system Venus, the Earth, and Mars represent respectively, youth, full growth, and old age. Venus, with its mountains rising dozens of miles into the atmosphere, is probably as yet unfitted for such existence as ours, but Mars must have passed through all terrestrial states and conditions.

Civilized existence rests on the development of the mechanical arts. The force of gravitation on Mars is only two-thirds of that on earth, hence all mechanical problems must have been much easier of solution. This is even more so of the electrical. The planet being much smaller, the contact between individuals and the mutual exchange of ideas must have been much quicker, and there are many other reasons why intellectual life should have been on that planet, phenomenal in its evolution.

To be sure, we have no absolute proof that Mars is inhabited. The straightness of the canals, which has been held out as a convincing indication to this effect, is not at all such. We can conclude with mathematical certitude that as a planet grows older and the mountains are leveled down, ultimately every river must flow in a geodetically straight line. Such straightening is already noticeable in some rivers of the earth.

But the whole arrangement of the so-called waterways, as pictured by Lowell, would seem to have been designed. Personally I base my faith on the feeble planetary electrical disturbances which I discovered in the summer of 1899, and which, according to my investigations, could not have originated from the sun, the moon, or Venus. Further study since has satisfied me that they must have emanated from Mars. All doubt in this regard will be soon dispelled.

To bring forth arguments why an attempt should be made to establish interplanetary communication would be a useless and ungrateful undertaking. If we had no other reason, it would be justified by the universal interest which it will command, and by the inspiring hopes and expectations to which it would give rise. I shall rather concentrate my efforts upon the examination of the plans proposed and the description of a method by which this seemingly impossible task can be readily accomplished.

The scheme of signaling by rays of light is old, and has been often discussed, perhaps, more by that eloquent and picturesque Frenchman, Camille Flammarion, than anybody else. Quite recently Prof. W. H. Pickering, as stated in several issues of the New York Times, has made a suggestion which deserves careful examination.

The total solar radiation falling on a terrestrial area perpendicular to the rays amounts to eighty-three foot pounds per square foot per second. This activity measured by the adopted standard is a little over fifteen one-thousandth of a horsepower. But only about 10 per cent of this whole is due to waves of light. These, however, are of widely different lengths, making it impossible to use all in the best advantage, and there are specific losses unavoidable in the use of mirrors, so that the power of sunlight reflected from them can scarcely exceed 5.5 foot pounds per square foot per second, or about one one-hundredth of a horse-power.

In view of this small activity, a reflecting surface of at least one-quarter million square feet should be provided for the experiment. This area, of course, should be circular to insure the greatest efficiency, and, with due regard to economy, it should be made up of mirrors rather small, such as to meet best the requirements of cheap manufacture.

The idea has been advanced by some experts that a small reflector would be as efficient as a large one. This is true in a degree, but holds good only in heliographic transmission to small distances when the area covered by the reflected beam is not vastly in excess of that of the mirror. In signaling to Mars, the effect would be exactly proportionate to the aggregate surface of the reflections. With an area of one-quarter million square feet the activity of the reflected sunlight, at the origin would be about 2,500 horse-power.

It scarcely need be stated that these mirrors would have to be ground and polished most carefully. To use ordinary commercial plates, as has been suggested, would be entirely out of the question, for at such an immense distance the imperfections of surface would fatally interfere with efficiency. Furthermore, expensive clock work would have to be employed to rotate the reflectors in the manner of heliostats, and provision would have to be made for protection against destructive atmospheric influence. It is extremely doubtful that so formidable an array of apparatus could be produced for $10,000,000, but this is a consideration of minor importance to this argument.

If the reflected rays were paralled and the heavenly bodies devoid of atmospheres, nothing would be simpler than signaling to Mars, for it is a truth accepted by physicists that a bundle of parallel rays, in vacuum, would illuminated an area with the same intensity, whether it be near or infinitely remote. In other words, there is no sensible loss in the transportation of radiant energy through interplanetary or vacuous space. This being the case, could we but penetrate the prison wall of the' atmosphere, we could clearly perceive the smallest object on the most distant star, so inconceivably tenuous, frictionless, rigid, and elastic is the medium pervading the universe.

The sun's rays are usually considered to be parallel, and are virtually so through a short trajectory, because of the immense distance of the luminary. But the radiations, coming from a distance of 93,000,000 miles, emanate from a sphere 865,000 miles in diameter, and, consequently, most of them will fall. on the mirrors at an angle less than 90 degrees, with the result of causing a corresponding divergence of the reflected rays. Owing to the equality of the angles of incidence and reflection, it follows that if Mars were at half the sun's distance, the rays reaching the planet would cover an area of about one-quarter of that of the solar disc, or in round numbers, 147,000,000,000 square miles, which is nearly 16,400,000, 000 times larger than that of the mirrors. This means that the intensity of the radiation received on Mars would be just that many times smaller.

To convey a definite idea, it may be stated that the light we get from the moon is 600,000 times feebler than that of the sun. Accordingly, even under these purely theoretical conditions the Pickering apparatus could do no more than produce an illumination 27,400,000 times feebler. than that of the full moon, or 1,000 times weaker than that of Venus.

The proceeding is based on the assumption that there is nothing in the path of the reflected rays except the tenuous medium filling all space. But the planets have atmospheres which absorb and refract. We see remote objects less distinctly, we perceive stars long after they have fallen below the horizon. This is due to absorption and refraction of the rays passing through the air.

While these effects cannot be exactly estimated it is certain that the atmosphere is the chief impediment to the study of the heavens.

By locating our observatories one mile above sea level the quantity of matter which the rays have to traverse on their way to the planet is reduced to one-third. But, as the air becomes less dense, there is comparatively little gain to be derived from greater elevation. What chance would there be that the reflected rays, reduced to an intensity far below that estimated above, would produce a visible signal on Mars? Though I do not deny this possibility, all evidence points to the contrary.

Lowell, a trained and restless observer, who has made the study of Mars his specialty, and is working under ideal conditions, has been so far unable to perceive a light effect of the magnitude such as the proposed signaling apparatus might produce there. Phobos, the smaller of the two satellites of Mars - from seven to 10 miles in diameter - can only be seen at short intervals when the planet is in opposition. The satellite presents to us an area of approximately fifty square miles, reflecting sunlight at least as well as ordinary earth, which has little over one-twelfth of the power of a mirror.

Stated otherwise, an equivalent effect at that distance would be produced by mirrors covering four square miles, which means two square miles of the same reflectors if located on earth, as it receives sunlight of twice the intensity. Now this is an area 222 times larger than that of the ten million dollar reflector, and yet Phobos is hardly perceptible. It is true that the observation of the satellite is rendered difficult by the glare of its mother planet. But this is offset by the fact that it is in vacuum and that its rays suffer little diminution through absorption and refraction of the earth's atmosphere.

What has been stated is thought sufficient to convince the reader that there is little to be expected from the plan under discussion. The idea naturally presents itself that mirrors might be manufactured which will reflect sunlight in parallel beams. For the time being this is a task beyond human power, but no one can set a limit to the future achievement of man.

Still more ineffective would be the attempt of signaling in the manner proposed by Dr. William R. Brooks and others, by artificial light, as the electric arc. In order to obtain a reflected light activity of 2,500 horsepower it would be necessary to install a power plant of not less than 75,000 horsepower, which, with its turbines, dynamos, parabolic reflectors and other paraphernalia, would probably cost more than $10,000,000. While this method would permit operation at favorable times, when the earth is nearer to, and has its dark side turned toward Mars, it has the disadvantage of involving the use of reflected rays necessarily more divergent than those of the sun, it being impossible to construct mirrors of the required perfection and without their use the rays would be scattered to such an extent that the effect would be much smaller.

Reflecting surfaces of great extent can be had readily. Prof. R. W. Wood makes the odd suggestion of using the white alkali desert of the southwest for the purpose. Prof. E. Doolittle advises the employment of large geometric figures. In my opinion none of these suggestions is feasible. The trouble is, that the earth itself is a reflector, not efficient, it is true, but what it lacks in this respect is more than made up by the immensity of its area. To convey a perceptible signal in this manner it might require as much as 100 square miles reflecting surface

But there is one method of putting ourselves in touch with other planets. Though not easy of execution, it is simple in principle. A circuit properly designed and arranged is connected with one of its ends to an insulated terminal at some height and with the other to earth. Inductively linked with it is another circuit in which electrical oscillations of great intensity are set up by means now familiar to electricians. This combination of apparatus is known as my wireless transmitter.

By careful attunement of the circuits the expert can produce a vibration of extraordinary power, but when certain artifices, which I have not yet described are resorted to the oscillation reaches transcending intensity. By this means, as told in my published technical records, I have passed a powerful current around the globe and attained activities of many millions of horsepower. Assuming only a rate of 15,000,000, readily obtainable, it is 6,000 times more than that produced by the Pickering's mirrors.

But, my method has other and still greater advantages. By its employment the electrician on Mars, instead of utilizing the energy received by a few thousand square feet of area, as in a parabolic reflector, is enabled to concentrate in his instrument the energy received by dozens of square miles, thus multiplying the effect many thousands of times. Nor is this all. By proper methods and devices he can magnify the received effect as many times again.

It is evident, then, that in my experiments in 1899 and 1900 I have already produced disturbances on Mars incomparably more powerful than could be attained by any light reflectors, however large.

Electrical science is now so far advanced that our ability of flashing a signal to a planet is experimentally demonstrated. The question is, when will humanity witness that great triumph. This is readily answered. The moment we obtain absolute evidences that an intelligent effort is being made in some other world to this effect, interplanetary transmission of intelligence can be considered an accomplished fact. A primitive understanding can be reached quickly without difficulty. A complete exchange of ideas is a greater problem, but susceptible of solution.

Nikola Tesla

What Science May Achieve this Year: New Mechanical Principle for Conservation of Energy

Denver Rocky Mountain News — Jan. 16, 1910

The spread of civilization may be likened to that of fire: First, a feeble spark, next a flickering flame, then a mighty blaze, ever increasing in speed and power. We are now in this last phase of development.

Human activity has become so widespread and intense that years count as centuries of progress. There is no more groping in the dark or accidentally stumbling upon discoveries. The results follow one another like the links of a chain. 'Such is the force of the accumulated knowledge and the insight into natural laws and phenomena that future events are clearly projected before our vision. To foretell what is coming would be no more than to draw logical conclusions, were it not for the difficulty in accurately fixing the time of accomplishment.

The practical success of an idea, irrespective of its inherent merit, is dependent on the attitude of the contemporaries. If timely it is quickly adopted; if not, it is apt to fare like a sprout lured out of the ground by warm sunshine, only to be injured and retarded in its growth by the succeeding frost. Another determining factor is the amount of change involved in its introduction. To meet with instant success an invention or discovery must come not only as a rational, but a welcome solution. The year 1910 will mark the advent of such an idea. It is a new mechanical principle.

Since the time of Archimedes certain elementary devices were known, which were finally reduced to two, the lever and the inclined plane. Another element is to be added to these, which will give rise to new conceptions and profoundly affect both the practical and theoretical science of mechanics.

This novel principle is capable of embodiment in all kinds of machinery. It will revolutionize the propulsion apparatus on vessels, the locomotive, passenger car and the automobile. It will give us a practical flying machine entirely different from those made heretofore in operation and control, swift, small and compact and so safe that a girl will be able to fly in it to school without the governess. But the greatest value of this improvement will be in its application in a field virtually unexplored and so vast that it will take decades before the ground is broken. It is the field of waste.

We build but to tear down. Most of our work and resource is squandered. Our onward march is marked by devastation. Everywhere there is an appalling loss of time, effort and life. A cheerless view, but true. A single example out of many will suffice for illustration.

The energy necessary to our comfort and safe existence is largely derived from coal. In this country alone nearly one million tons of the life-sustaining material are daily extracted from the bowels of the earth with pain and sacrifice. This is about seven hundred tons per minute, representing a theoretical activity of, say, four hundred and fifty-million horsepower. But only a small percentage of this is usefully applied.

In heating, most of the precious energy escapes through the flue. The chimneys of New York City puff out into the air several million horsepower. In the use of coal for power purposes, we hardly capture more than 10 percent. The exhaust of engines carries off more energy than obtained from live steam.

In many modern plants the power has been actually doubled by obviating this waste, but the machinery employed is cumbersome and expensive. The manufacture of light is in a barbarous state of imperfection, and this may also be said of many industrial processes. Consider just one case, the manufacture of iron and steel.

America produces approximately 30,000,000 tons of pig iron per year. Each ton of iron requires about one and a half tons of coal, hence, in providing the iron market, 70,000,000 tons of coal per annum, or 133 tons per minute, are consumed. In the manufacture of coke a ton of coal yields, roughly, 10,000 cubic feet of gas of a mean heating capacity of 600 heat units per cubic foot.

Bearing in mind that 133 tons are used per minute, the total heat units developed in that time would be 798,000,000, the mechanical equivalent of which is about 19,000,000 horsepower. By the use of the new principle 7,000,000 horsepower might be rendered available. A furnace of 200 tons produces approximately 17,000 cubic feet of gas per minute of heat value of 100 units, corresponding to a theoretical performance of 40,000 horsepower, of which not less than 13,000 might be utilized in the improved apparatus referred to. The power derived by this method from all blast furnaces in the United States would be considerably above 5,000,000 horsepower.

The preceding figures, which are conservative, show that it would be possible to obtain 12,000,000 horsepower merely from the waste gasses in the iron and steel manufacture. The value of this power, fairly estimated, is $180,000,000 per annum, and it must be made worth much more by systematic exploitation.

A part of the power could be advantageously employed for operating the blowers, rollers and other indispensable machinery and supplying electricity for smelting, steel making and other purposes. The bulk might be used in the manufacture of nitrates, aluminum, carbides and ice. The production of nitrates would be particularly valuable from the point of view of national economy. Assuming that 5,000,000 horsepower were apportioned for that purpose, the annual yield would be not less than 10,000,000 tons of concentrated nitric compound, adequate to fertilizing 40,000,000 acres of land. A great encouragement would be given to agriculture and the condition of the steel and iron workers ameliorated by offering to them a fertilizer at a reduced rate, thus enabling them to cultivate their farms with exceptional profit. Other conveniences and necessities, as light, power, ice and ozonized water could be similarly offered and numerous other improvements, both to the advantage of capital and labor, carried out.

To appreciate the above it should be borne in mind that the iron and steel industry is one of the best regulated in the world. In many other fields the waste is even greater. For example, in the operation of steam railroads, not less than 98 per cent of the total energy of coal burned is lost. An enormous saving could be effected by replacing the present apparatus with new gas turbines and other improved devices for transmitting and storing mechanical energy. A study of this subject will convince that for the time being, at least, there is more opportunity for invention in the utilization of waste than in the opening up of new resources.

N. Tesla

Mr. Tesla on the Future

Modern Electrics — May, 1912

On Tesla Day, at the Northwest Electric Show, held at Minneapolis, Minn., March 16th to 23rd, Mr. Tesla sent, through Archbishop Ireland, the following message to the people of the Twin Cities and the Northwest:

New York, N. Y., March 18, 1912. His Grace, The Most Reverend Archbishop Ireland:

I bespeak your Grace's far-famed eloquence in voicing sentiments and ideas to which I can give but feeble expression. May the exposition prove a success befitting the cities of magical growth, the courage and energy of western enterprise, a credit to its organization, a lasting benefit to the communities and the world through its lessons and stimulating influence as a bewildering, unforgettable record of the triumphant progress of the art. Great as are the past achievements, the future holds out more glorious promise. We are getting an insight into the essence of things; our means and methods are being refined, a new and specialized race is developing with knowledge deep and precise, with greater powers and keener perceptions. Mysterious as ever before, nature yields her precious secrets more readily and the spirit of man asserts its mastery over the physical universe. The day is not distant when the very planet which gave him birth will tremble at the sound of his voice; he will make the sun his slave, harness the inexhaustible and terribly intense energy of microcosmic movement; cause atoms to combine in predetermined forms; he will draw the mighty ocean from its bed, transport it through the air and create lakes and rivers at will; he will command the wild elements; he will push on and on from great to greater deeds until with his intelligence and force he will reach out to spheres beyond the terrestrial.

I am your Grace's most obedient servant.

Nikola Tesla

Tesla and Marconi

New York Sun — May 22

To the Editor of The Sun - Sir: The reports contained in The Sun and other journals regarding the issue of a recent wireless patent suit are of a nature to create an erroneous impression. Two of the patents mentioned, namely, Nos. 11,9j3 and 609,154, granted respectively to William Marconi and Sir Oliver Lodge, are of no importance, but another patent of the former expert, dated June 28, 1904, contains arrangements on which I obtained full protection more than three years before and which are essential to the successful practice of the wireless art at any considerable distance.

My patents bear the numbers 645,576 and 649,621 and were secured through Kerr, Page & Cooper, attorneys for the General Electric and Westinghouse companies. The apparatus described by me comprises four circuits peculiarly arranged and carefully attuned so as to secure the greatest possible flow of electrical energy through them. The generator is a transformer of my invention and the oscillations employed are of a kind which are now known in technical literature as the Tesla currents. Every one of these elements, even to the last detail, is contained in the Marconi patent which was involved in the suit, and its use constitutes an infringement of all the fundamental features of my wireless system.

Nikola Tesla
New York, March 21, 1914.

Nikola Tesla Tells of Country's War Problems
New York Herald — April 15, 1917

The conquest of elements, annihilation of distance in the transmission of force and numerous other revolutionary advances have brought us face to face with problems new and unforeseen. To meet these is an imperative necessity rendered especially pressing through the struggle which is now being waged between nations on a stupendous scale unprecedented in history.

This country, finding it impossible to remain an inactive witness of medieval barbarism and disregard of sacred rights, has taken up arms in a spirit broad and impartial and in the interest of humanity and peace. Its participation will be absolutely decisive as regards the final result, but those who expect a speedy termination of the conflict should undeceive themselves.

War, however complex, is essentially a mechanical process, and, in conformity with a universal principle, its duration must be proportionate to the masses set in motion. The truth of this law is borne out by previous records, from which it may be calculated that, barring conditions entirely out of the ordinary, the period should be from five to six years.

Great freedom of institutions, such as we are privileged to enjoy, is not conducive to safety. Militarism is objectionable, but a certain amount of organized discipline is indispensable to a healthy national body. Fortunately, the recognition of this fact has not come too late, for there is no immediate danger, as alarmists would make us believe. The geographical position of this country, its vast resources and wealth, the energy and superior intelligence of its people, make it virtually unconquerable.

There is no nation to attack us that would not be ultimately defeated in the attempt. But events of the last three years have shown that a combination of many inimical powers is possible, and for such an emergency the United States is wholly unprepared. The first efforts must therefore be devoted to the perfection of the best plea for national protection. This idea has taken hold of the minds of people and great results may be expected from its creative imagination fired by this occasion, such as may in a larger measure recompense for the awful wastage of war.

While the chief reliance in this perilous situation must be placed on the army and navy, it is of the greatest importance to provide a big fleet of airplanes and dirigibles for quick movement and observation; also a great number of small high speed craft capable of fulfilling various vital duties as carriers and instruments of defense. These, together with the wireless, will be very effective against the U-boat, of which the cunning and scientific enemy has made a formidable weapon, threatening to paralyze the commerce of the world.

As the first expedient for breaking the submarine blockade, the scheme of employing hundreds of small vessels, advanced by Mr. W. Denman, chairman of the United States Shipping Board, is a most excellent one, which cannot fail to succeed. Another measure which will considerably reduce the toll is to use every possible means for driving the lurking enemy far out into the sea, thus extending the distance at which he must operate and thereby lessening his chances. But a perfect apparatus for revealing his presence is what is most needed at this moment.

A number of devices, magnetic, electric, electro-magnetic or mechanical, more or less known, are available for this purpose. In my own experience it was demonstrated that the small packet boat is capable of affecting a sensitive magnetic indicator at a distance of a few miles. But this effect can be nullified in several ways. With a different form of wireless instrument devised by me some years ago it was found practicable to locate a body of metallic ore below the ground, and it seems that a submarine could be similarly detected.

Sound waves may also be resorted to, but they cannot be depended upon. Another method is that of reflection, which might be rendered practicable, though it is handicapped by experimental difficulties well nigh insuperable. In the present state of the art the wireless principle is the most promising of all, and there is no doubt that it will be applied with telling effect. But we must be prepared for the advent of a large armored submarine of great cruising radius, speed and destructive power which will have to be combated in other ways.

For the time being no effort should be spared to develop aerial machines and motor boats. The effectiveness of these can be largely increased by the use of a turbine, which has been repeatedly referred to in the *Herald* and is ideally suited for such purposes on account of its extreme lightness, reversibility and other mechanical features.

N. Tesla

Tesla Answers Mr. Manierre and Further Explains the Axial Rotation of the Moon

New York Tribune — Feb. 23, 1919

Sirs:

In your article of February 2, Mr. Charles E. Manierre, commenting upon my article in "The Electrical Experimenter" for February, which appeared in The Tribune of January 26, suggests that I give a definition of axial rotation.

I intended to be explicit on this point, as may be judged from the following quotation: "The unfailing test of the spinning of a mass is, however, the existence of energy of motion. The moon is not possessed of such vis viva." By this I meant that "axial rotation" is not simply "rotation upon an axis" as nonchalantly defined in dictionaries, but is circular motion in the true physical sense - that is, one in which half the product of the mass with the square of velocity is a definite and positive quantity.

The moon is a nearly spherical body, of a radius of about 1,081.5 miles, from which I calculate its volume to be approximately 5,300,216,300 cubic miles. Since its mean density is 3.27, one cubic foot of material composing it weighs close to 205 pounds. Accordingly, the total weight of the satellite is about 79,969,000,000, 000,000,000,000 and its mass 2,483,500,000,000,000,000 terrestrial short tons. Assuming that the moon does physically rotate upon its axis, it performs one revolution in 27 days 7 hours 43 minutes and 11 seconds, or 2,360,591 seconds. If, in conformity with mathematical principles, we imagine the entire mass concentrated at a distance from the center equal to two-fifths of the radius, then the calculated rotational velocity is 3.04 feet per second, at which the globe would contain 11,474,000,000,000,000,000 short foot tons of energy, sufficient to run 1,000,000, 000 horsepower for a period of 1,323 years. Now, I say that there is not enough energy in the moon to run a delicate watch.

In astronomical treatises usually the argument is advanced that "if the lunar globe did not turn upon its axis it would expose all parts to terrestrial view. As only a little over one-half is visible it must rotate." But this inference is erroneous, for it admits of one alternative. There are an infinite number of axes besides its own on each of which the moon might turn and still exhibit the same peculiarity.

I have stated in my article that the moon rotates about an axis, passing through the center of the earth, which is not strictly true, but does not vitiate the conclusions I have drawn. It is well known, of course, that the two bodies revolve around a common center of gravity which is at a distance of a little over 2,899 miles from the earth's center.

Another mistake in books on astronomy is made in considering this motion equivalent to that of a weight whirled on a string or in a sling. In the first place, there is an essential difference between these two devices though involving the same mechanical principle. If a metal ball attached to a string is whirled around and the latter breaks an axial rotation of the missile results which is definitely related in magnitude and direction to the motion preceding. By way of illustration: If the ball is whirled on the string clockwise, ten times a second, then when it flies off it will rotate on its axis twenty times a second, likewise in the direction of the clock. Quite different are the conditions when the ball is thrown from a sling. In this case a much more rapid rotation is imparted to it in the opposite sense. There is not true analogy to these in the motion of the moon. If the gravitational string, as it were, would snap, the satellite would go off in a tangent without the slightest swerving or rotation, for there is no momentum about the axis and, consequently, no tendency whatever to spinning motion.

Mr. Manierre is mistaken in his surmise as to what would happen if the earth were suddenly eliminated. Let us suppose that this would occur at the instant when the moon is in opposition. Then it would continue on its elliptical path around the sun, presenting to it steadily the face which was always exposed to the earth. If, on the other hand, the latter would disappear at the moment of conjunction, the moon would gradually swing around through 180 degrees and, after a number of oscillations, revolve again with the same face to the sun. In either case there would be no periodic changes, but eternal day and night, respectively, on the sides turned toward and away from the luminary.

Nikola Tesla

Made in the USA
Lexington, KY
05 December 2010